Current Trends on
Lanthanide Glasses and Materials

by
Sooraj Hussain Nandyala

This book entitled 'Current Trends on Lanthanide Glasses and Materials' is an innovative monograph covering the latest developments in lanthanide doped glasses and phosphors materials. The book aims to explain the basic mechanism of phosphor materials, and the luminescence behaviour of glasses doped with certain lanthanide ions. It also describes how to plot colors in a CIE chromaticity diagram.

The book will be of use for senior researchers, materials scientists, chemists, physicists, engineers, as well as research students to gain knowledge on current developments of these materials.

Cover page Figure Description: Transparent fluorophosphate glasses are among interesting amorphous materials to incorporate a large amount of rare earth ions and providing a long excited state lifetime of such emitters. The cover presents the CIE diagram of Eu-Tb co-doped fluorophosphate glasses, manifesting their color tunability as the function of the rare earth concentration and applied excitation wavelength.

Current Trends on Lanthanide Glasses and Materials

Edited by

Sooraj Hussain Nandyala

School of Metallurgy and Materials, University of Birmingham, Edgbaston, Birmingham B15 2TT, UK

Published by **Materials Research Forum LLC**
Millersville, PA 17551, USA

Published as part of the book series
Materials Research Foundations
Volume 8 (2017)
ISSN 2471-8890 (Print)
ISSN 2471-8904 (Online)

Print ISBN 978-1-945291-14-2
ePDF ISBN 978-1-945291-15-9

Distributed worldwide by

Materials Research Forum LLC
105 Springdale Lane
Millersville, PA 17551
USA
http://www.mrforum.com

Manufactured in the United State of America
10 9 8 7 6 5 4 3 2 1

Table of Contents

Dedicated to

!... In the memory of ...... my beloved late teacher ..... !

Srinivasa Buddhudu, Professsor and Vice Principal of Sri Venkateswara University College, Tirupati, India

(August 15, 1955 - July 30, 2014)

On behalf of All Former PhD and M. Phil Students of Professor Srinivasa Buddhudu
&
Dr. Sooraj H. Nandyala, University of Birmingham, United Kingdom.

Former PhD student

Preface

The present monograph entitled "Current Trends on Lanthanide Glasses and Materials" highlight the novel findings of Lanthanide materials. Chapter one starts by reviewing the fundamentals of glasses, presenting the advantages of fluorophosphate glasses which follow by spectroscopic formulation of the rare earth ions. Several examples of new fluorophosphate glasses doped with rare earth ions are presented and their spectroscopic and structural properties are discussed. Co-doped lanthanide ions doped fluorophosphate glasses are nominated as good candidates as white light generating or color tunable solid state materials.

Chapter two describes the Erbium (Er^{3+}) doped phosphate glasses which are striking materials due to their broadband emission at around 1.55 µm, thanks to various transitions from 4I13/2 state to 4I15/2 manifolds. More important, incorporation of metallic nanoparticles in Er3+ doped phosphate glasses have been introduced as an interesting method to enhance their optical properties for broadband applications.

Chapter three summarizes the recent progress on the suitability of lanthanum-contained iron borophosphate for nuclear waste immobilization from the aspect of structure, chemical durability, irradiation stability, thermal properties, etc.

Recent studies on the crystallization behaviors and crystallization kinetics of the cerium containing iron borophosphate glasses/glass-ceramics were summarized in chapter four.

Li_2O-MO-B_2O_3 glasses have high thermal expansion coefficient and have high mechanical strength and are transparent to both UV and visible regions. These glasses can easily be prepared and were considered as the good materials for applications in battery sealing and for enamel paints. Though some studies on certain physical properties of lithium borate glasses are available in the literature, the detailed systematic investigations on electrical, optical properties and thermoluminescence properties of Li_2O-MO-B_2O_3: Nd^{3+} and Sm^{3+} glasses are not available. Therefore, chapter five studies may help for considering the applications of these glasses in radiation dosimeters.

The effect of compositional changes on the structure, thermal and X-ray absorption properties of fluoro-phosphate glasses have been reported in the chapter six.

Tellurite glasses are considered as the best materials for optical components such as IR domes, optical filters, modulators, memories and laser windows in view of their high transparency in the far infrared region and for their high density and refractive index. Therefore, a relationship between the structural modifications and luminescence efficiencies of ZnF_2-MO-TeO_2 glasses doped with Ho^{3+} and Er^{3+} ions have been reported in the chapter seven.

The development of novel and potential ultraviolet (UV)/blue LEDs based on wideband gap semiconductor such as GaN led to considerable progress in the field of solid state lighting. Therefore, in the final chapter, it is intended to explain the luminescence and energy transfer phenomena of lanthanide ions doped in phosphor materials.

Eventually, the edior would like to thank the publisher for its help and co-operation.

Sooraj Hussain Nandyala

School of Metallurgy and Materials, University of Birmingham, Edgbaston, Birmingham UK

CHAPTER 1

Tunable and white light generation in lanthanide doped novel fluorophosphate glasses

M. Reza Dousti

Physics Institute, Federal University of Alagoas, Macéio, AL, 57072-970, Brazil

Abstract

Fluorophosphate glasses have attracted increasing attention due to their superior structural and optical properties such as mediated phonon energy, good chemical and thermal stability, wide transmittance window, and providing a long lifetime for the radiative emissions of the rare earth ions. Therefore, the aim of this chapter is to present a general portrait of the current achievements and challenges on the preparation and properties of the rare earth ions doped fluorophosphate glasses. The chapter starts by reviewing the fundamentals of glasses, presenting the advantages of fluorophosphate glasses which are followed by spectroscopic formulation of the rare earth ions. Finally, several examples of the new fluorophosphate glasses doped with rare earth ions are presented and their spectroscopic and structural properties are discussed. Co-doped lanthanide ions doped fluorophosphate glasses are nominated as good candidates as white light generating or color tunable solid state materials.

Keywords

Fluorophosphate Glasses, Co-Doping, Rare Earth Ions, White Light Generation, Color Tunable, Spectroscopic Properties

Contents

1.   Introduction

1.1   The art of glasses

Recently, searching for modern materials with longer lifetime and higher efficiency has attracted greater attention. A good examples are solar cells. One can name a series of novel materials having a high research interest rate such as polymers, ceramics, glasses and new light sources based on lasers, amplifiers in networking, Li-ion batteries, etc. The development of high efficiency new materials is promising to aid the "green energy" agenda. Glasses are one of the most exciting materials that are found in many daily used objects and applications such as drinking cups, wine bottles, mirrors, electric lamps, window glass, decorative objects, optical fibers, and army as well as nuclear waste forms. The glasses can be synthesized in various colors and desirable shape and size.

The most common methodology to prepare a glass is to rapidly cool down the molten of one or more glass-formers. The ashes of volcanos contain aluminum, silicon, sodium, potassium, calcium and iron and therefore natural cooled residuals of avalanched volcano's ashes are indeed the first ever glass pieces on the earth.

The term, "glass", in general is attributed to any non-crystalline material (amorphous) whose structure lacks any long- or short-range order, although it shows a glass transition. Thus, various materials such as ionic melts, metallic alloys, aqueous solutions, polymers etc. are known as glasses. Glass is classified as fragile materials, and its color could be adjusted by selection of glass host, modifiers and/or dopants such as transition metals and rare earth ions. Vitrification is a process by which a heated substance (molten) could be transformed to a glass through a rapid cooling down to its glass transition temperature. Such fast chilling is necessary to prevent the formation of any crystalline phase.

Currently, glass-ceramics have attracted a large attention. They can be formed by controlling the nucleation process through one or two step heat-treatment on the parent glass; the mother glass is exposed to heat in order to nucleate the crystals and then at elevated temperatures, the nuclei start to grow. In principal, nucleating agents have to be added to speed-up and control the ceramization. Glass-ceramics are among polycrystalline materials showing both properties of glasses and ceramics. Therefore, its structure benefits from both amorphous and crystalline phases and results in a low porosity materials with high toughness, high temperature stability, good chemical durability, ion conductivity, high break down voltage, better optical properties etc. Glass-ceramics can be mass-produced and their nanostructure can be designed according to desired properties [1].

The history of glass preparation, art and science of the glasses is not specified to a region, and glass blowers have been known to move from country to country due to several geo-political issues at their time. Figure 1 shows the chronogram map of the glass development around world as highlighted in detail elsewhere [2,3].

In the new millennium, new glasses are designed for diverse applications as decorative objects, daily life facilities as well as scientific tools such as lens, solid state lasers, goggles, artificial bones and teeth, etc. The improved understanding of amorphous nature of glasses, their local structure, preparation methods, cutting and polishing techniques etc. high-tech glasses are designed for micro- and telescopes, glass fibers and high power lasers [3].

Calcium phosphate, silicate and borosilicate glasses have been used as nuclear glass waste forms, as well as artificial bones for bio-techniques [4,5]. Currently, solders are prepared from zinc phosphate and lead borate glasses [6–8]. Phosphate and fluro-phosphate glasses were applied as laser hosts [9]. Chalcogenide glasses are used as memory panels and switching materials [10]. Halide glasses are substituted for silica fibers due to their excellent transparency [11,12]. Tellurite glasses attracted large attention because of their significant optical, thermal and physical properties. Physical and optical properties of glasses depend on the elements of choice in their glass composition which may have one or two glass-formers to combine various properties, such as borosilicate and borofluorite, fluorophosphate, etc. Recently, oxyfluoride (fluorophosphate) glasses have attracted a large attention for optical properties because they provide comparatively low phonon energies and high chemical and mechanical stabilities [13].

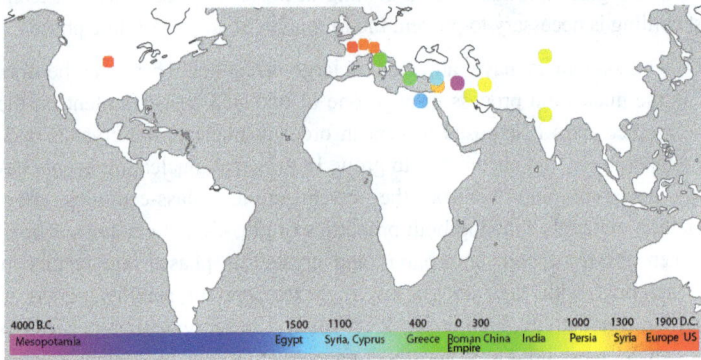

Figure 1. Centers of glass-blowers during the past 6000 years. The original map is adopted from www.outline-world-map.com @2009. Data are taken from [3].

In order to utilize glasses for various applications such as solid state lasers, army tools, medical facilities etc. Excellent understanding of the correlation between the structure and property of glasses is of high demand. This necessity results in the emergence of wide research topics on glasses that can be better observed by recording the number of scientific publications on each particular field. Figure 2 presents the number of publication on "phosphate glasses", 'fluoride glass" and "fluorophosphate glasses" indexed by SCOPUS database since 1940 up to the end of December 2015 [14]. An exponential increase in number of publications on phosphate and fluorophosphate glasses in general shows this increasing need. This is also interesting to note that a number of publications on "fluoride glasses" was almost constant during the years 1993 to 1998, while the number of SCOPUS-index publications on "fluorophosphate glass" fluctuated during the last decade. Mauro and Zanotto [15] have literally reviewed the increment in the number of publication on glass technology. Their data was also taken from the SCOPUS database and such analysis concluded that China and the United States are the most prolific countries in glasses in terms of number of publications, and Journal of Non-

Crystalline Solids is the frontier journal by publishing a large number of "glass-related" journal articles.

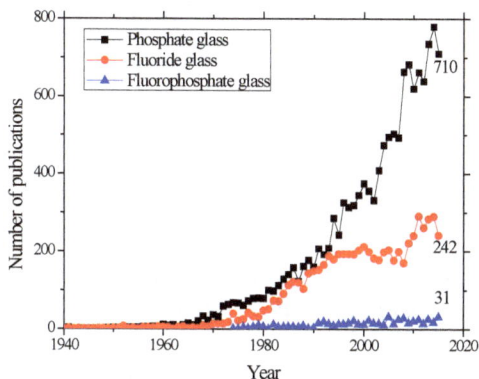

Figure 2. Number of Scopus-index publications on phosphate, fluoride and fluorophosphate glasses. Data taken from Scopus on February 3rd, 2016 [14].

1.2 Phosphate glasses

Phosphate glasses are well-known and widely studied glass formers since they showed high capacity of vitrification and are capable to dissolve high amounts of other glass formers and modifiers [16]. These specific properties are due to easy formation of covalent linear, bidimensional and tridimensional networks in phosphor oxides, which in turn increases the viscosity in the liquid state and favors the glass formation. The incorporation of network modifiers such as alkaline oxides in P_2O_5 results in the progressive depolimerization of the covalent tridimensional network formed by the PO_4 tetrahedra, which consist of three bridging P-O-P bonds. When depolymerized, phosphate chains end with terminal phosphates units (P-O terminal bonds) and such a breakage at the network chain indicates a progressive decrease of viscosity. Hence, the addition of large levels of modifiers deforms the covalent network and isolated ionic units with no ability to vitrify could be formed [17]. Figure 3 shows the different tetrahedral sites of a phosphate glass.

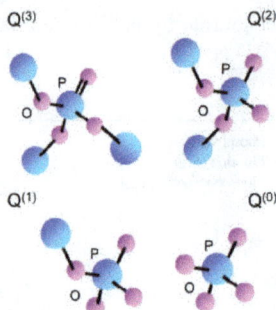

Figure 3. Probable tetrahedral sites in a phosphate glass; Q^3: ultraphosphate, Q^2: metaphosphate, Q^1: pyrophosphate, and Q^0: orthophosphate. Figure is adopted from [18].

Phosphate glasses are also known candidates for optical applications, since they have a higher refractive index than silicate glasses and show good transparency in the UV-Vis region. The unique disadvantage of using phosphate glasses as luminescence materials is their high phonon energy (~1200 cm^{-1}) which could cause emission quenching due to energy transfer from fluorophores to OH groups. On the other hand, phosphate glasses possess other beneficial properties - such as high thermal expansion coefficient - which makes them promising candidates for diverse applications.

1.3   Fluoride glasses

Fluoride glasses have been in the center of attention of the glass and ceramic research due to their good transparency in UV-edge and very low phonon energy (500–600 cm^{-1} [19]) which extends their high transmittance in the infrared region. Such small maximum phonon energy results in the quantum efficiencies and excellent optical properties of the rare earth doped fluoride glasses. However, the physical and chemical properties of fluoride glasses - e.g. mechanical strength, chemical durability and thermal stability - are generally less than oxide glasses such as phosphate glasses. These properties are important for developing rare earth-doped optical fibers [19]. Still, the fluoride glasses could be stabilized by addition of network modifiers and other glass formers such as scandium polyphosphate [20].

One of the major disadvantages of the fluoride glasses is high loss of fluorine during the melting procedure which is hazardous for furnace compartments. Moreover, fluoride glasses lack enough mechanical, thermal and chemical stability when compared to oxide glasses.

The advantage and disadvantage of both the phosphate and fluoride glasses are given in Table 1. Looking at those properties, it is clear that the addition of oxide glasses such as phosphate in the fluorine glass composition could significantly strength the composition, while keeping a high transparency in a wide range of the spectra.

1.4   Fluorophosphates glasses

As already discussed above, fluorine glasses could not be prepared with precise stoichiometry molar percentage due to the high losses of fluorine during the melting process which may also affects the performance of the furnaces. Moreover, the final fluorine glass has low chemical curability, although they benefit from their low-phonon cut-off energy characteristics, which drastically improves the radiative properties of the doped rare earth ions. However, if a small amount of phosphates is added to the fluorine molten, the glass formability can be greatly improved thanks to the formation of PO covalent bonds at the intermediate of the fluoride glass structure [21].

The infrared transmission of the fluorophosphate glasses is less than fluorine glass and that increases their maximum phonon energy. However, the concentration of OH groups is still lower than oxide glasses due to the presence of fluorine [22]. According to the study, fluorophosphate glasses possess better chemical and physical properties than fluoride glasses and near to phosphate glasses [23]. Rather than this, rare earth ions doped PbF_2-aluminosilicate glasses or LaF_3-alumino-silicate-based glass-ceramics have been also prepared, where the lanthanide ions substitute in the PbF_2 and LaF_3 lattices of the crystallized component [24,25]. Preparation of glasses with high rare earth concentration is the current challenge to develop high-power lasers. Therefore, new rare earth doped fluorophosphate glasses are proposed and their local structure and optical performance are characterized [21,26–29]. Production of high homogenous fluorophosphate glasses is reviewed elsewhere [30]. However, understanding the glass structure, radiative properties of rare earth ions as dopants, and correlation between structure and properties of the fluorophosphate glasses still requires further devotion [22,29].

These glasses have important technological applications in optics. Beryllium-free commercial fluorophosphate glasses have been available since the 1950's. The formal fluorophosphate glasses for optical applications could not be produced in large scale because of the low viscosity of its molten and high tendency to crystallize. Later, other fluorophosphate glasses with low amount of oxygen were prepared, where new methodologies aimed to produce glasses with higher optical quality, and gradually fluorophosphate with improved transmittance in the entire UV to IR region were developed [31].

Fluorophosphate glasses are structurally interesting since wide range of fluorides could be introduced in their network. The glass formation region in fluorophosphate glasses is wide and large, and various glass modifiers could be adjusted to obtain the suitable characteristics. For example, fluorophosphate glasses can be prepared by addition of (79-85) LiF, or (72-81) NaF, or (66-78) KF, or (0-35) SrF_2, or (66-74) BaF_2, or (0-25) AlF_3 (all amounts are given in mol%) into the $Al(PO_3)_3$ [32].

One of the importance of fluoride glasses as an optics media is the existence of low phonon energies that facilitates the non-linear phenomenon such as upconversion. The upconversion mechanism in phosphate glasses is very weak due to the high phonon energies of the lattice ($\sim$1200-1300 cm^{-1}), provided by the stretching vibration of the P-O bond. Such drawback could be overcome by introduction of fluoride contents which possess lower phonon energy ($\sim$200-600 cm^{-1}). In this case, when media doped with rare earth ions, the multi-phonon decay rates decreases and quantum yield increases.

Fluorophosphate glasses are a better candidate than fluoride glasses since they benefit from higher chemical durability, easy preparation techniques in air atmosphere, low crystallization tendency, and lower effect on melting crucible and furnace [22]. Although phosphate network could be to some extend detrimental for optical performance. Therefore, it is of important to control the amount of the phosphate content to optimize the optical properties [33].

The thermal properties of the AlF_3 - $NaPO_3$ and AlF_3 – CaF_2 – $NaPO_3$ glasses have been studied. It is shown that the crystallization is more distinct in high fluorine content glasses. The thermal dilatometry coefficient increases by increasing the amount of fluorine in these studied fluorophosphate glasses [34]. Other studies showed an increment in the mechanical and thermal properties of phosphate glasses by addition of fluoride [35–37]. The density, optical path change with temperature (1.4x10^{-6} $^{\circ}$C^{-1}) and thermal expansion coefficients (14.x10^{-6} $^{\circ}$C^{-1}) of fluorophosphates glasses is higher than silicate and phosphate commercial laser glasses [26]. The glass transition temperature and resistivity to devitrification can be varied in fluorophosphate glasses by the addition/removal of some modifying components [38]. Compared to silicate glasses, the damage threshold of laser glasses are low for fluorophosphate glasses due to their low thermal endurance. Such low damage threshold in fluorophosphate glasses is an impediment property of this class of non-crystalline materials which arose from their low phosphoric oxide content, resulting in a fragile network [39]. The infrared absorption spectra of the fluorophosphate glasses formed with high concentration of $Al(PO_3)_3$ is similar to crystalline $Al(PO_3)_3$ which has a metaphosphate structure, while with low $Al(PO_3)_3$ content, is consist of pyrophosphate groups. The decrease in the amount of $Al(PO_3)_3$ which indeed is an increases in fluorine content results in formation of a

$(PO_3F)^{2-}$ group [32]. The nonlinear indices of fluorophosphate glasses are lower than those of silicates or phosphates used in present laser fusion systems because the hyperpolarizability of the fluorine anion is smaller than that of oxygen. The optical properties of the rare earth doped fluorophosphate glasses are intermediate between the spectral behaviour of pure phosphate and fluoride glasses [40].

Table 1. Summary of the optical and durability performance of the phosphate, fluoride and fluorophosphate glasses.

	Phosphate glasses	Fluoride glasses
Advantages	1. 1. High mechanical stability 2. Good rare earth dispersitivity	2. 1. Low hygroscopicity 2. Low phonon energy (500 cm^{-1})
Disadvantages	1. 1. High hygroscopicity 2. High phonon energy (1200 cm^{-1})	1. Mechanically fragile
Fluorophosphate glasses		
1. High chemical and mechanical stability		
2. High transmittance in UV edge		
3. Good rare earth solubility		
4. Lower hygroscopicity		

1.5   Spectroscopic properties of rare earth ions doped glasses

Rare earth ions doped lanthanide glasses are interesting materials due to both the properties of the glasses and various radiative emissions from lanthanides in a wide range of spectra. For example, glasses are good hosts for rare earth ions thanks to their high rare earth solubility, wide transmittance window, high linear and non-linear refractive indices and good chemical and thermal stability. On the other hand, rare earth ions are capable to emit sharp lines with relatively longer lifetimes (in case of $4f^N$ to $4f^N$ transitions) than typical fluorophores. They could also provide transitions with shorter lifetimes due to the $4f^N$ to $4f^{N-1}5d$ transitions [41]. On the other hand, energy transfer among the rare earth ions could results in particular transitions. Such energy transfer processes could assist in various mechanism such as upconversion emissions or may be detrimental due to ion-ion or ion-host interactions which quenches the radiative emissions. The rare earth ions doped glasses and glass-ceramics are promising materials for solid state lasers and broadband fiber amplifiers which work at visible and inferred regions, respectively lasers [42].

Basically, the energy level diagram of the lanthanide ions is symbolized according to the Russell-Saunders definition ($^{2S+1}L_J$), where L, S and J are the orbital angular, spin angular and total angular momentums, respectively. As provided in quantum mechanism, the energy levels of each rare earth ions results from various interactions among the nuclei, electrons and surrounding medium, which could be theoretically written as the sum of the corresponding Hamiltonians [43]

$$H = H_{free-ion} + H_{"Crytal-field"}$$

where

$$H_{free-ion} = -\frac{\hbar^2}{2}\sum_{i=1}^{N}\nabla_i^2 - \sum_{i=1}^{N}\frac{Z^*e^2}{r_i} + \sum_{i<j}^{N}\frac{e^2}{r_{ij}} + \sum_{i=1}^{N}\zeta(r_i)\vec{s}_i.\vec{l}_i + negligible\ terms.$$

and N is the total number of electrons in the 4f shell, Z^* is the effective charge of nuclei, which takes into account the inner electrons and nuclei, $\zeta(r_i)$, s_i and l_i are the efficiency of the spin-orbit coupling, and spin and orbit angular momentum, respectively. Therefore, all the kinetic energy, Coulomb interaction, mutual Coulomb repulsion and spin-orbit interaction of the 4f electrons interfere to define the energy levels. The degeneracy of the 4fN electronic configuration results in the broadening of energy levels due to the mutual Coulomb repulsion and spin-obit interactions. The Hamiltonian of the crystal field is around 100 times weaker than electrostatic and spin-orbit interactions, due to strong shielding by the electrons of 5s and 5p solids [43], but it still is able to induce "Stark splitting" at the asymmetric crystal field of solids. The nature of the forbidden intra 4fN electric-diploe transitions changes in presence of a small amount of excited opposite parities into the mixture of 4fN.

The above-mentioned Hamiltonian could be also affected in the presence of the electromagnetic field (excitation light) and interactions between neighboring ions as

$$H = H_{f-ion} + H_{CF} + V_{EM} + V_{ion-ion}$$

where V_{EM} and $V_{ion-ion}$ represent the Hamiltonian of interaction of light by ion, and the interactions between the ions. V_{EM} is responsible for the absorption transitions, when the frequency of the incoming magnetic field is in resonance or near-resonance with transition between (often) the ground state and various excited states of the rare earth ions.

One of the drawbacks of the incorporation of more rare earth ions in the glassy host matrix is the detrimental energy transfer among these ions which results in "concentration quenching" of radiative transitions which reduces the luminescence intensity [44,45]. In order to overcome such phenomena, several proposals are given in the literature. The main ideas deal with incorporation of new species which could increase the quantum efficacy either by energy transfer or plasmonic effects, although modification of crystal field by changing the glass host and the network modifiers are also examined to a great extent [2,46].

The glasses doped with rare earth ions have various technological applications such as optical fibers, amplifiers, solid state lasers, sensor, upconvertors etc. During the last decade, many works have been invented and further research are ongoing to develop the superior candidates with higher efficiency and durability. A summary of the achievements on rare earth doped commercially available technologies are listed in Table 2.

Table 2. Early developments in rare earth doped materials for technological applications.

Year	Inventor/developer	Property	Working wavelength (nm)
1961	Snitzer [47]	first Nd^{3+} doped fiber pumped by solid state flash lamp	1061 and 1062
1964	Koester, & Snitzer [48]	near single-Nd^{3+}-doped fiber laser	~ 1060
1966	Kao, & Hockham [49]	theory of propagation in core-clad fibers	
1972	Sandoe et al. [50]	first phosphate glass containing Er^{3+} ions	1530-1560
1973	Stone, & Burrus [51]	Nd^{3+} fibers (CW laser)	800
1986	Mears et al. [52]	first tunable and Q-switched fiber laser	1528-1542 and 1544-1555
1990	-	First Transatlantic fiber optic TAT-8 cables	-
1997	Svendsen	optical network WDM system	-

1.5.1 Ion-ion interactions and non-linear processes

The interactions between two active ions located in short vicinity of each other could be classified as the energy transfer processes through which sensitization by ion S can affect the neighboring ion A. Such processes are illustrated schematically in Fig. 4. In this model, the sensitizer ion is in its excited state and the activated ion is in the ground state. The energy transfers could be either between two levels at resonant energies or in two non-resonant energy states where lattice phonons assist to provide the energy difference.

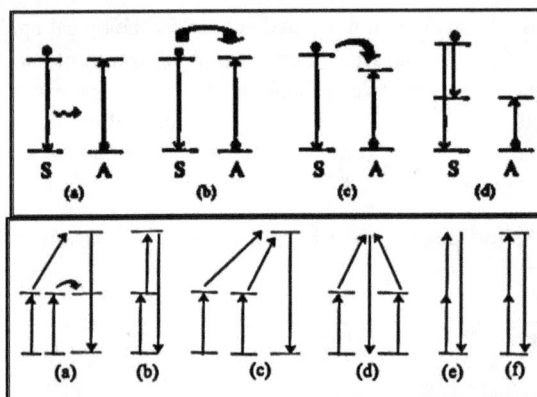

Figure 4: (Above) Energy transfer processes from a sensitizer (S) to an activator (A). Resonant radiative transfer (a), resonant energy transfer (b), Energy transfer assisted by phonons (c), and example of quenching of the fluorescence of S by energy transfer to A (d). (Below) Schematic energy transfer mechanism for different two photon upconversion processes. APTE effect in YF_3:Yb:Er (a), 2-steps absorption in SrF_2:Er (b), cooperative sensitization in YF_3:Yb:Tb (c), cooperative luminescence in Yb:PO_4 (d), Second harmonic generation in KDP (e), and 2-photon absorption excitation in CaF_2:Eu^{2+} (f).

Various energy transfer processes have been introduced by Auzel [53]. He considered that energy transfer may occur even when the activator ion is in its excited state. Then, he also discussed cooperative luminesce, second harmonic generation, 2-photon excitation and 2-step absorption, as well as cooperative sensitization, as shown in Fig. 4. Another probable interactions between neighboring ions is the cross-relaxation processes which is the reverse path of the cooperative upconversion process. The 2-step excitation (or excited state absorption) is based on the absorption of a photon by an ion in the excited metastable level.

The upconversion process is a Dexter energy transfer [54] in which the ions at metastable excited state act as the activators. This process is a "short-range" energy transfer, and the rate of the process decreases exponentially by increasing the ion-ion distances. Generally, the energy transfers are more probable at high concentration of the rare earth ions [55], and in case that its rate is higher than other existing radiative and non-radiative transitions from the metastable level. The rate of the excited state absorption increases by increasing the intensity of the pump.

1.5.2 Principals of Judd-Ofelt theory

Judd-Ofelt theory [56,57] describes the radiative transition of the rare earth ions using either the optical absorption or crystal spectra of the rare earth ions [58]. The theory was developed by two researchers independently and simultaneously with similar assumptions and results. The theory is commonly used to evaluate the optical efficiency of the rare earth doped materials by finding the absorption oscillator strength, radiative probabilities and branching ratio of every transition and intrinsic lifetime of each excited state. The absorption and radiative emission properties are correlated with a set of parameters called as Judd-Ofelt intensity parameters Ω_i (i=2,4 and 6).

Although alternative methods has been proposed to determine the Judd-Ofelt intensity parameters from the collected spectrum of the rare earth ions [59,60], the main approach is to calculate the experimental oscillator strengths (f_{exp}) of each transition in the absorption spectrum and compare them to the calculated formula of oscillator strengths which is derived theoretically as

$$f_{exp} = \frac{4.318 \times 10^{-9}}{N} \int \alpha(\omega)d\omega$$

$$f_{cal}[(S,L)J; (S',L')J'] = \frac{8\pi^2 mc}{3h\lambda e^2(2J+1)}\left[\frac{(n^2+2)^2}{9n}S_{ed} + n^3 S_{md}\right]$$

where α and N correspond to the absorption coefficient (cm^{-1}) and number of active ions (mol.L^{-1}), and e, m, c, h and λ have their common definitions in physics. Here, $|(S,L)J >$ and $|(S',L')J' >$ shows the quantum numbers of ground states and excited states, respectively, for an electric-dipole transition. Moreover, S_{ed} and S_{md} are the corresponding line-strengths of electric-dipole and magnetic-dipole contributions at each wavelength, respectively, and defined as [61]

$$S_{ed} = e^2 \sum_{t=2,4,6} \Omega_t |\langle (S,L)J\|U^{(t)}\|(S',L')J'\rangle|^2$$

$$S_{md} = \frac{e^2\hbar^2}{4m^2c^2} \left| \langle (S,L)J \| \vec{L} + 2\vec{S} \| (S',L')J' \rangle \right|^2$$

By and large, a least-square fitting is the most common method to compare the experimental and calculated oscillator strengths through which three intensity parameters (Ω_i) could be derived. It is worthy to note that the reduced matrix elements are independent from choice of host and could be find elsewhere [62]. The Judd-Ofelt intensity parameters of the Er^{3+} ions in different hosts are listed in Table 3. Finally, the transition probabilities A, branching ratio β and radiative lifetime τ are given as

$$A_{J \to J'} = \frac{64\pi^4}{3h(2J+1)\lambda^3} [\chi S_{ed} + n^3 S_{md}]$$

$$\beta = \frac{A_{J \to J'}}{\sum_{J'} A_{J \to J'}}$$

$$\tau = \frac{1}{\sum_{J'} A_{J \to J'}}$$

Table 3. Three Judd-Ofelt intensity parameters for the Er^{3+} ions in various glasses ($\times 10^{-20}$ cm^2).

Glass type	Ω_2	Ω_4	Ω_6
Silicate [63]	5.59	1.42	0.87
Phosphate [63]	4.67	1.37	0.77
Tellurite [63]	5.34	1.75	0.94
Germanate [63]	5.72	0.91	0.32
Fluoride [63]	2.91	1.27	1.11
Borate [64]	4.11	1.45	1.42
Phospho-tellurite [65]	4.25	1.50	0.43
Fluorophosphate [22]	4.36	2.29	1.36
Fluorozirconate ZBLA [66]	2.54	1.39	0.97
Kigre phosphate [67]	6.28	1.03	1.39

One of the important parameters of the rare earth doped materials is the excited state lifetime which defines the transition rate and capability to obtain the population inversion and intense amplification in optical amplifier. In general, the Judd-Ofelt theory could define the intrinsic lifetime of each excited state which differ from the experimental values due to the presence of the non-radiative decays and energy transfer processes. Such non-radiative decays which are mainly known as multi-phonon relaxations result from the interaction of the excited state ions and the phonons of the network. For example, consider the broadband emission of Er^{3+}-doped glasses which emits at around 1530 nm due to $^4I_{13/2} \rightarrow {}^4I_{15/2}$ transition. In the phosphate glasses, due to the large phonon energy and proximity of the energy state of the OH group with $^4I_{13/2}$ excited state of Er^{3+} ions, large probability of non-radiative decays cause a huge decreases in the lifetime and intensity of 1530 nm band, while such properties could be improved in a low phonon energy glass host such as tellurite (~ 700 cm^{-1}) or fluoride glasses. Therefore, fluorophosphate glasses are good candidates which could benefits from both better optical performance of fluorides and chemical and thermal stability of phosphates. The Judd-Ofelt analysis and spectroscopic properties of the fluorophosphate glass with composition $75NaPO_3$-$24BaF_2$-$1LnF_3$ (where Ln= Pr^{3+}, Nd^{3+}, Sm^{3+}, Eu^{3+}, Tb^{3+}, Dy^{3+}, Ho^{3+}, Er^{3+} and Tm^{3+}) are studied elsewhere [68].

2.   Summary of the ongoing research

As mentioned in Section 1.1, fluorophosphates glasses have attracted a large attention due to the combined structural and optical properties of phosphate and fluoride glasses which results in mechanically stronger, optically transparent and structurally non-hygroscopic glass pieces. Thanks to the superior properties of such glass compositions, different research groups have started to develop studies on structural and photophysical properties of fluorophosphates glasses containing various modifier and intermediate components, suitable as hosts for rare earth ions and metallic nanoparticles.

For example, Raaben et al. [69] reported the optical properties and radiative probabilities of the Nd^{3+} ion doped fluorophosphates glasses having composition 15 $Ba(PO_3)_2$ – 40 AlF_3 – 5 MgO – 20 CaF_2 – 20 SrF_2 (mol%). They observed that the fluorophosphates glasses possess a wider transmittance window in the UV edge than common silicate and phosphate glasses, while the absorption of Nd^{3+} ions in the red region of spectra is relatively intense. A detailed Judd-Ofelt analysis has been done on both the absorption and luminescence spectra of glasses doped with 0.1 to 3.0 mol% of Nd_2O_3, concluding an absolute quantum yield close to unity. Deutschbein et al. [70] investigated the phosphate and fluorophosphates glasses doped with Nd^{3+} ions. They reported that Nd^{3+} doped phosphate glasses have a great laser effect, while adding fluorine content to them could decrease the non-linear refractive index, which makes them suitable candidates for the nuclear fusion experiments. The lifetime of the emission band at 1060 nm ($^4F_{3/2} \rightarrow {}^4I_{11/2}$) in Nd^{3+} ions in fluorophosphates glass (510 and 460 µs for LG-810 and LG-800 glasses) is higher than phosphate glasses (290 µs for LG-700 glass). The low dispersion (Abbe number ~ 90) and low refractive index (~ 1.45) of fluorophosphates glasses result in a low non-linear refractive index and a long excited state lifetime of the rare earth ions. Such properties are interesting for high power laser applications [26].

On the other hand, fluorophosphate glasses containing heavy metals could possess a higher refractive index, e.g., n~1.7 for $Pb(PO_3)_2$ - NaF/CaF_2 and 1.6 for $Ba(PO_3)_2$ – NaF/CaF_2 glass compositions, which respectively have densities of about 4.6 and 3.7 g.cm^{-3} [71], which is higher than silicate (~2.54 g.cm^{-3} for ED-2) and phosphate (2.85 g.cm^{-3} for LHG-8) glasses. Moreover, the structural and thermal studies have been investigated on the $RF–RF_2–AlF_3–Al(PO_3)_3$ fluorophosphate glasses where R=Li, Na, K, Mg, Ca, Sr, and Ba. The infrared-Fourier absorption spectroscopy and its complement, Raman spectroscopy on these glass series revealed that the sum of P-O-P and O-P-O bonds increases by increasing the alkali and alkaline earth fluoride components, which results in stronger glass network. Therefore, the glass resistance to the nucleation and crystallization processes is enhanced as probed by the improved difference between glass crystallization and transition temperatures (ΔT) [37].

The rare earth doped fluorophosphate glasses showed promising emission properties such as high cross-sections and long lifetimes. For example, the excited state lifetime of the $^6P_{5/2}$ level of Gd^{3+} ions doped (Al-Y-Gd)F_3 – (Ca-Sr-Ba)F_2 fluorophosphate glasses is equal to 9.2 ms [72]. The optical properties and Judd-Ofelt analysis of Pr^{3+} ions doped fluorophosphate glasses having composition 50(NaPO$_3$)$_6$ - 10Zn$_3$(PO$_4$)$_2$ - 10BaF$_2$ - 9AlF$_3$ – xAF – (15-x)BF - 1PrF$_3$ (where x=5 and 10 mol% and A and B=Na/K/Li) are studied [73]. The glass show and average refractive index of about 1.488 and density of about 2.43-2.57 gr.cm^{-3}. The branching ratio of the $^3P_0 \rightarrow {}^3H_4$ transition in these glasses is more than 72%, which indicates a great laser action probability for the emission line at 480 nm, when excited at 445 nm.

The luminescence spectra of Eu^{3+} ions doped alumino-fluorophosphate glasses with compositions range of (50-x-y)AlF$_3$ - xAlPO$_4$ – yEuF$_3$ – 30CaF$_2$ – 20BaF$_2$ are studied [74]. It is concluded that the asymmetry and covalancy around the Eu^{3+} ions increases by increasing the AlPO$_4$ content. Further, it is observed that the P-O bonds with high phonon energy is preferentially coordinated to the Eu^{3+} ions, and that increase by addition of AlPO$_4$. Therefore, the intensity ratio between the luminescence bands at 610 nm ($^5D_0 \rightarrow {}^7F_1$) and 590 nm ($^5D_0 \rightarrow {}^7F_2$) increases by increasing the phosphate concentration, when excited at 394 nm. The same behavior in the intensity ratio of the two latter transitions in

Eu^{3+} doped tungsten sodium phosphate glasses is observed, where the intensity ratio decreases from 3.83 to 2.21 for 0 and 40 mol% of PbF_2 in this glasses [75]. Furthermore, the lifetime of the 610 nm emission of Eu^{3+} ions increase from 0.92 ms to 2.27 ms by increasing the PbF2 content from 0 to 40 mol%.

The luminescence dynamics of Tm^{3+} doped fluorophosphate glasses $50(NaPO_3)_4$ – $18BaF_2$ – $10ZnF_2$ – $20RF$ – $2TmF_2$ mol%, where R=Li, Na, K and Li-Na, Na-K and K-Li) is studied by Reddy et al. [76]. It is observed that the glass containing LiF possesses the highest density ($g.cm^{-3}$), while the glasses co-maintaining KF and LiF have lower density ($g.cm^{-3}$). The refractive index of this glass system is around 1.48. The Judd-Ofelt analysis has been carried out and it is concluded that glasses containing NaF have stronger optical intensities. The upconversion emission of Tm^{3+}/Yb^{3+} ions co-doped fluorophosphate glasses is also investigated and a strong emission line at 480 nm is observed [77]. It is concluded that the addition of Yb^{3+} ions to the singly doped Tm^{3+} glasses increases the upconversion intensity up to 200 times.

Upconversion emission in Ho^{3+} doped fluorophosphate glasses is also reported in which blue, green, and red upconversion emissions are observed using glass composition as $7Ba(PO_3)_2$ - $32AlF_3$ - $30CaF_2$ - $18SrF_2$ - $13MgF_2$ [33]. The upconversion emission bands are observed at 491, 543 and 658 nm and interpreted by excited state absorption, energy transfer and cross-relaxation mechanism. The infrared-Fourier reflection spectroscopy is used to identify the low phonon energy characteristics of proposed fluorophosphate glasses ($\sim$600 cm^{-1}) which yields in low multi-phonon decay rates and improvement in the upconversion efficiency.

The broadband emission of Er^{3+} ions in fluorophosphate glasses is investigated, as well, due to their vital importance for the optical fiber technology. The first Er^{3+} doped fluorophosphate glass fiber amplifier is reported to have potential for amplifying an 8 channel WDM signal in the wavelength rage of about 1532-1560 nm. Low signal noise (< 6dB) and high gain ($\sim$8.2dB) are observed which lie in the promising range for such parameters [78]. Moreover, Er^{3+} doped $8NaH_2PO_4$ - $11Al(PO_3)_3$ - $10MgHPO_4$ - $22LiF$ - $17AlF_3$ - $11SrF_2$ - $8CaF_3$ - $9MgF_2$ -$4BaF_2$ glasses show a broadband in the 1400-1650 nm spectral region whose lifetime is measured to be 7.6-8.4 ms, and the FWHM of about 53 nm. Such broadening is higher in compare to silicate and phosphate glasses, while lower than tellurite and bismuth based glasses [79]. The difference between the crystallization temperature and glass transition temperature is about 168°C which is larger than tellurite glasses ($\sim$110°C) and comparable to bismuth based glasses (123-170°C). Therefore, this glass composition could be used in the fiber drawing process to decrease the optical scattering of optical fibers. Moreover, fluorophosphate glasses with composition also show strong thermal stability around 230°C [80].

The emission cross-section of 1.53 μm emission of Er^{3+} doped fluorophosphate glasses is reported to be higher than those of silicate, germinate and tellurite glasses [36]. Recently, new fluorophosphate glasses doped with Yb^{3+} ions are reported which possess low concentration of hydroxyl group (~3-7 ppm) and high emission cross section (0.7-0.8 pm^2) at around 1000 nm. The lifetime of the latter band (from Yb^{3+}) ions in proposed fluorophosphate glasses could be as long as 2.67 ms [81]. Such high emission cross section and long lifetime are reported in various Yb^{3+} doped fluorophosphate glasses *for high-power* generation [35,82]. Blue, green and 0.8 μm Tm^{3+}, Ho^{3+} doped upconversion laser glasses, sensitized by Yb^{3+} is also reported by Peng et al. [19].

The "Faraday effect" is also reported in some fluorophosphate glasses, generally doped with Tb^{3+} ions. For example, The "Farady effect" is reported to be higher in borate and fluoride glasses than fluorophsosphate glasses doped with Tb^{3+} ions, where the magnitude of the Verdet constant (V_c) follows the order, borate glass > fluoride glasses > fluorophosphate glass [83]. The small V_c in the studied glasses is due to their small number of magnetic ions per unit volume compare to borate glasses. However, a patent has been registered [84] on the "Faraday rotation glass", where a fluorophosphates glass was prepared. Such glass composition showed a large Verdet constant and a low non-linear refractive index.

Recently, new fluorophosphate glasses with compositions BaF_2 - SrF_2 - $Al(PO_3)_3$ –AlF_3 doped with Eu^{3+} ,Tb^{3+} ,Er^{3+} and/or Yb^{3+} ions are prepared in the Laboratory of Advanced Functional Materials (LEMAF) of the Institute of Physics at University of São Paulo. The aim of the research group is to prepare fluorophosphate glasses with excellent transparency, and good thermal stability and chemical durability, and to correlate their structure and properties; mainly the optical properties by the aim of photoluminescence spectroscopy and excited state lifetime measurements and structural properties by EPR or NMR spectroscopy and thermal analysis (DSC). Promising results have been presented up to the publication date of this chapter [22,29,85], and further research are developing in this group. These studies are relatively new and very little was known on the structure of such type of glasses, medium-range order of these glasses and the probable sites of the rare-earth ions. The studied glass compositions and their corresponding labels are given in Table 4, and pictures of the samples are given in Figure 5.

Table 4. Compositions (mol%) and labels of studied glasses in References [22,29,85].

Glass label	BaF$_2$	SrF$_2$	Al(PO$_3$)$_3$	AlF$_3$	YF$_3$	ErF$_3$	YbF$_3$	TbF$_3$	EuF$_3$
FP20	25	25	20	10	20	-	-	-	-
FP15	25	25	15	15	20	-	-	-	-
FP10	25	25	10	20	20	-	-	-	-
FP05	25	25	05	25	20	-	-	-	-
FP20-RE025	25	25	20	10	19.75	0.25		-	-
FP15-RE025	25	25	15	15	19.75	0.25		-	-
FP20-RE05	25	25	20	10	19.50	0.50		-	-
FP15-RE05	25	25	15	15	19.50	0.50		-	-
FP20-RE1	25	25	20	10	19	1		-	-
FP15-RE1	25	25	15	15	19	1		-	-
FP20-RE2	25	25	20	10	18	2		-	-
FP15-RE2	25	25	15	15	18	2		-	-
FP20-RE3	25	25	20	10	17	3		-	-
FP15-RE3	25	25	15	15	17	3		-	-
FP20-RE4	25	25	20	10	16	4		-	-
FP15-RE4	25	25	15	15	16	4		-	-
FP20-RE5	25	25	20	10	15	5		-	-
FP15-RE5	25	25	15	15	15	5		-	-
FP20-Er025Yb4	25	25	20	10	15.75	0.25	4	-	-
FP15-Er025Yb4	25	25	15	15	15.75	0.25	4	-	-
FP20-Er1Yb4	25	25	20	10	15	1	4	-	-
FP15-Er1Yb4	25	25	15	15	15	1	4	-	-
FP20-Eu01	25	25	20	10	19.90	-	-	-	0.10
FP20-Eu025	25	25	20	10	19.75	-	-	-	0.25
FP20-Eu05	25	25	20	10	19.50	-	-	-	0.50
FP20-Eu1	25	25	20	10	19	-	-	-	1.00
FP20-Eu2	25	25	20	10	18	-	-	-	2.00
FP20-Eu3	25	25	20	10	17	-	-	-	3.00

FP20-Eu4	25	25	20	10	16	-	-	-	4.00
FP20-Tb01	25	25	20	10	19.90	-	-	0.10	-
FP20-Tb025	25	25	20	10	19.75	-	-	0.25	-
FP20-Tb05	25	25	20	10	19.50	-	-	0.50	-
FP20-Tb1	25	25	20	10	19	-	-	1.00	-
FP20-Tb2	25	25	20	10	18	-	-	2.00	-
FP20-Tb3	25	25	20	10	17	-	-	3.00	-
FP20-Tb4	25	25	20	10	16	-	-	4.00	-
FP20-EuTb03	25	25	20	10	19.70	-	-	0.15	0.15
FP20-EuTb05	25	25	20	10	19.50	-	-	0.25	0.25
FP20-EuTb1	25	25	20	10	19	-	-	0.50	0.50
FP20-EuTb1.5	25	25	20	10	18.50	-	-	0.75	0.75

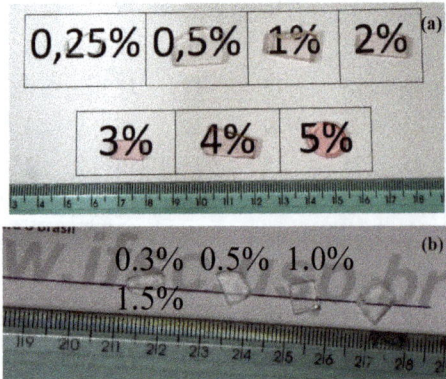

Figure 5. Photographs of the Er^{3+}-doped (a) and Tb^{3+}/Eu^{3+} co-doped (b) fluorophosphate glasses. The numbers give the molar concentration of dopants.

3.   Rare earth doped new fluorophosphate glasses

3.1   Structural properties

By and large, addition of phosphate oxide content to the fluorine glasses increases the glass transition temperature and increases the thermal stability of glasses. While the density of the samples decreases by increasing the phosphate oxide content and the refractive index increases [23]. However, such properties could be optimized by the addition of network modifiers and finding the appropriate O/F ratio [20]. Moreover, the dissolution rate of fluorine glasses can be decreased by the addition of phosphate oxide, which in turn increases the surface hardness of the obtained fluorophosphate glasses [23].

Raman spectra of the un-doped fluorophosphate glasses (with compositions given in Table 5) are shown in Figure 6. Several bands are observed in the 400-1400 cm^{-1} energy region which could be associated to the alkali earth-fluorine bond (500 cm^{-1}), Al-F vibrations (570 cm^{-1}), symmetric stretching of P-O-P linkages (around 660 and 750 cm^{-1}, respectively for Q^2 and Q^1 tetrahedral species) and also PO_4 and P-O stretching around 1000 and 1080 cm^{-1} which includes the Q^0 and Q^1 species. The band of metaphosphate units occurs around 1200 cm^{-1} which is due to stretching vibration of P-O linkages. The P-F stretching vibration has energy of about 800-900 cm^{-1}, however such band is not observed in the Raman spectra of the studied fluorophosphate glasses [22].

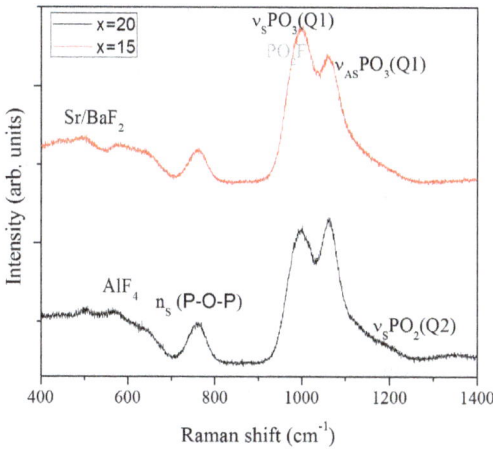

Figure 6. Raman spectra of the fluorophosphate glasses having compositions 25 BaF_2 – 25 SrF_2 – 20 YF_3 – (30-x) $Al(PO_3)_3$ – x AlF_3 (mol%). Original spectra are given [22]

Table 5. Theoretical and experimental wavenumbers of the probable structural groups in fluorophosphate glasses.

Structural group	Theoretically calculated		Experimental wavenumber (cm^{-1})		
	Stretching force constant (N m^{-1})	Wavenumber (cm^{-1})	This work[*]	BCaAlFP[**]	Literature[***]
O–P–O	892	1691	~ 1200	1280	1240-1310[a]
O–P–F	887	1521	not observed	1140	1120-1210[b]
F–P–F	882	1440	not observed	980	750-1050[c]
[*] [22], [**] Ba(PO$_3$)$_2$ - 2MgF$_2$ - 10CaF$_2$ - AlF$_3$ [86], [***] Ref. [a][87], [b][88] , [c][89]					

The network connectivity at Na-P-O-F and Na-Al-P-O-F glasses has been studied by Brow et al [90]. The ^{19}F NME spectra reveal strong F-P chains. Such bonding causes a little shift on the NMR signal of ^{31}P, showing that fluorine substitutes with bridging oxygen of the phosphate tetrahedral. For the glass system of Na-Al-P-O-F composition, only octahedral sites of Al-sites are detected by using [17]Al NMR spectroscopy, while ^{19}F NMR and ^{31}P spectra revealed the F-Al bonding and Q1-sites (with one or more Al-sites at next-nearest neighbors), respectively. It's been also observed, that re-melting the glass in NH$_4$HF$_2$ results in the formation of a greater number of F-P bonds [90].

The NMR studies of the xMnF$_2$–(x)MnF$_2$–(x–80)NaPO$_3$–(20)ZnF$_2$ glasses show long single bond –O-PO$_2$- chains could be broken by non-bridging fluorine [91]. This is confirmed by the results of infrared absorption spectra of these glasses, where by addition of fluorine content, bridging oxygens becomes non-bridging and a non-bridging fluoride locates in a PO3F tetrahedron. Therefore, the Q^2 chains start to be shorter and transform into Q^1 species. Pure NaPO$_3$ glass consists of almost only Q^2 tetrahedra. The addition of only 20 mol% of ZnF$_2$ results in formation of Q^1 tetrahedra (up to 40%). With further addition of ZnF$_2$, almost all the Q^2 tetrahedra sites disappear, indicative of formation of P$_2$(O, F)$_7$ pyrophosphate groups which are connected to the MF$_6$ octahedra through the vitreous network [91].

The NMR spectroscopy technique is used to quantify the amount of fluorine in the proposed fluorophosphate glasses. It is very common to lose and/or replace some amount of fluorine by oxide species during the melting procedure. The experimental method is given elsewhere in details [22], and the results showed that about 80% of the fluorine is preserved in the final glass network. The weight loss of the glasses is negligible, therefore, most of the fluorine loss could be attributed to absorption of some amount of oxygen or water by the glass network during preparation.

The REDOR NMR spectroscopy technique allows determining the ratio between the number of F (N_F) and P (N_P) atoms for the first coordination sphere of the Al. It has been observed that by increasing the amount of AlF_3, in expense of $Al(PO_3)_3$, N_P/N_F ratio and total number of P-O-Al linkages are decreased. The number of P atoms in the coordination of Al also decreases from 4.4 to 1.0 by increasing the x from 10 to 25 mol% (see glass compositions in Table Y), while it is 6 for the $Al(PO_3)_3$ fluorine-free system [29].

The thermal analysis of the proposed fluorophosphate glasses in the fluorophosphate glass composition of Table 4 indicates that the glass transition temperature of these samples is at a maximum for 15 mol% of AlF_3 content which decreases by further addition of this component. The composition dependent glass transition temperature values are illustrated in Figure 7.

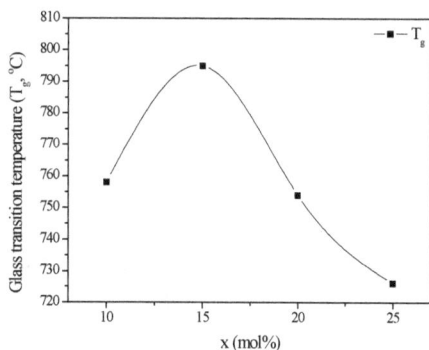

Figure 7. Compositional dependent of the glass transition temperature in the glass system as $30-x)Al(PO_3)_3 - xAlF_3 - 25\ BaF_2 - 25SrF_2 - 20\ YF_3$. Data taken from [29].

3.2   Optical properties

The UV-Vis-IR transparency of the optical glasses is an important factor which could determine the applicability of the proposed composition in various cases. By and large, fluoride and fluorophosphate glasses show good transparency in the UV spectral region. The UV-edge (cut-off) of the glass composition as $Ba(PO_3)_2$–MgF_2–CaF_2–AlF_3 shifts towards lower energy by increasing the fluorine content up to 36% [92].

The optical properties of the new fluorophosphate glasses such as $0.4MgF_2$-$0.4BaF_2$-$0.1Ba (PO_3)_2$-$0.1Al(PO_3)_3$ doped with rare earth ions was studied and it was found that the refractive index of these glasses increases by increasing the concentration of rare earth ions. These glasses show a relatively large Abbe number which is independent of rare earth concentration [93].

The fluorophosphate glasses containing high amount of tungsten oxide are also prepared and a large two-photon absorption coefficients has been observed when excited by a 660 nm nanosecond pulsed laser controlled [94]. Such un-doped glasses have promising applications as optical limiters controllers [94], and are promising upconvertors when doped with Tm^{3+} ions [95].

3.2.1 Near-infrared and upconversion emissions

Due to the good transparency and mechanical strength of the rare earth doped fluorophosphate glasses, they are suitable candidates for solid state lasers, optical fibers, sensors and upconvertors, taking into account the longer lifetime provided by F environment. For example, Er^{3+} ions could optically activate the fluorophosphate glasses thanks to their green (~540 nm), broadband (~ 1.53 µm) or upconversion emissions. High emission cross section (1.86-3.14 x 10-20 cm2) and long intrinsic lifetime of $^4I_{13/2}$ excited state (10.39-11.03 ms) of Er^{3+} ions doped fluorophosphate glasses are reported [96]. Such new materials are excellent candidates for developing broadband optical amplifiers and compact fiber lasers. However, the quantum efficiency of such system could be still improved by Yb^{3+} ions co-doping, since this ion has a larger absorption cross-section which could facilitate the upconversion emission and increase the lifetime of the $^4I_{13/2}$ excited state of the Er^{3+} ions. The refractive index of the proposed alumina fluorophosphate Glasses (composition given in Table 4) is around 1.537 at 546.07 nm [22]. In general, the refractive index of most of the fluorophosphate glasses lies in this range [93,97], except for the heavy metal fluorophosphate glasses such as lead-fluorophosphates which have a refractive index of about 1.7 [71].

Er^{3+}, Yb^{3+}, singly- and Er^{3+}-Yb^{3+} co-doped fluorophosphate glasses are also studied [22]. Such glasses show good transparency and excellent rare earth solubility. The near-infrared and upconversion emissions of these glass sets are investigated, as presented in Figure 8. The broadband of the Er^{3+} ions centered at about 1533 nm has a FWHM of about 86.3 nm and a lifetime of bout 8-9 ms. The Yb^{3+} ions also show a near-infrared broadband at about 42.3 nm, with a lifetime of about 1.5 ms. It is important to note that the values of FWHM depend strongly on the thickness of the sample, dopant concentration and excitation light power [98]. Therefore, these values are not compared to other materials in this chapter. The Judd-Ofelt analysis on the glasses is performed and three intensity parameters (Ωi= i=2,4 and 6) are listed in Table 3. The physical concept of such intensity parameters has been controversial in the literature. However, Ω_2 is mainly attributed to the covalent character of the rare earth ions and the site symmetry of former ions. Therefore, in the fluorophosphate glasses the values of Ω_2 is lower than in phosphate glasses. It is due to the covalent nature of such oxide glasses, which decreases by addition of less covalent fluorine atoms.

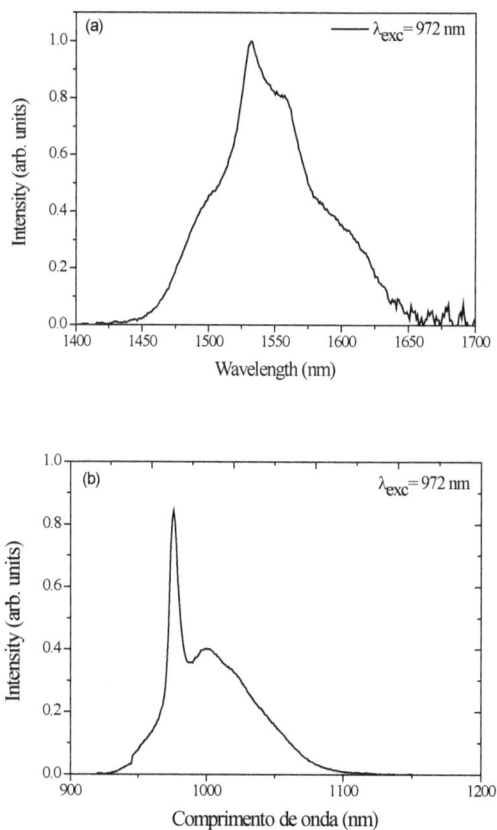

Figure 8. Broadband emission of 2mol% Er^{3+} (a) and Yb^{3+} (b) singly-doped fluorophosphate glass.

The quantum efficiency of the $^4I_{13/2}$ excited state of Er^{3+} ions and $^2F_{7/2}$ excited states of Yb^{3+} ions could be calculated by giving the ration of experimental lifetime to the intrinsic lifetime values. The intrinsic lifetime values are those obtained from Judd-Ofelt properties. The lifetime values and quantum efficiency of the singly and co-doped Er^{3+}/Yb^{3+} fluorophosphate glasses are listed in Table 6. The obtained long experimental lifetime could be related to various aspects. The main factor is the incorporation of fluorine which decreases the volume fraction of oxygen and OH group which are detrimental for the broadband luminescence of the rare earth ions. Another factor is the sample thickness which can seriously alter the experimental lifetime values due to the so-called "reabsorption" process. Therefore, the lifetime has to be recalculated, which is beyond the scope of this chapter and details can be found elsewhere [98,99]. However, luminescence quantum efficiency above 100 percent is obviously the result of a re-absorption phenomenon in thick samples. In the co-doped samples an increase in the lifetime of the $^4I_{13/2}$ state of Er^{3+} and decrease in the $^2F_{7/2}$ state of Yb^{3+} ions are indicatives of the energy transfer from Yb^{3+} ions to Er^{3+} ions upon excitation at 980 nm.

Table 6. Lifetime and quantum efficiency of the singly and co-doped Er^{3+}/Yb^{3+} fluorophosphate glasses [22].

Singly and co-doped glasses	RE= Er^{3+} ions		RE= Yb^{3+} ions	
	τ_{exp} (ms)	η (%)	τ_{exp} (ms)	η (%)
20RE0.25	9.14	76	-	-
15RE0.25	8.55	76	-	-
20RE0.5	9.83	82	-	-
15RE0.5	7.76	69	-	-
20RE1	9.02	75	1.70	131
15RE1	8.56	76	1.45	131
20RE2	8.36	69	-	-
15RE2	9.13	82	-	-
20RE3	7.63	64	1.69	130
15RE3	8.83	79	1.50	135
20RE4	7.92	66	-	-
15RE4	8.14	73	-	-
20RE5	5.18	43	1.58	120

15RE5	6.42	57	1.53	138
20Er0.25Yb4	8.98	75	0.88	68
15Er0.25Yb4	8.18	73	0.90	81
20Er1Yb4	9.02	75	0.90	69
15Er1Yb4	8.83	79	0.91	82
20Er2Yb4	10.16	85	0.49	38
15Er2Yb4	9.22	83	0.48	43

3.2.2 Tunable excitation and white light generation

Recently, tunable light source has attracted great attention. Rare earth ions doped glasses are interesting materials which could provide various emission lines in the visible region. The incorporation of two or more rare earth ions result in colorful glasses which can emit at various wavelengths and general special combination of colors, provide different color by changing the excitation wavelength or excitation pulse duration [85,100]. On the other hand, combination of generated colors could results in white light which has attracted attention due to importance of the development in white light emitting diodes (WLEDs). The advantages of WLEDs such as energy savings, longer lifetime, environmentally clean and color tunability has made them superior light source which could substitute the standard fluorescent tubes. Since the rendering index of the commercial WLESs is not efficient, it is suggested to use a near-ultraviolet GaN-based chip working at around 350-400 nm, and emitting at red, green and blue spectra region. The emitting materials could be selected according to the dopants, optical, and thermal performance as well as chemical durability.

Three different systems of rare earth doped glasses will be presented here as examples of white light generating and tunable excitation materials. The schematic luminescence spectra of each system are presented in Figure 9. Fluorophosphate triply-doped with Tm^{3+}:Tb^{3+}:Mn^{2+} ions can emits at around 450 nm (Tm^{3+}:$^1D_2 \rightarrow {}^3F_4$), around 543 nm ($Tb^{3+}$:$^5D_4 \rightarrow {}^7F_5$) and a broader band in entire 575-700 nm region due to Mn^{2+} ions, when excited at 360 nm wavelength. The color combination results in a white light whose CIE coordinates is (0.3527, 0.3146) [101].

Moreover, fluorophosphate glasses doped with Tb^{3+}:Eu^{3+} ions are introduced as good candidates for tunable light generation whose color can be switched by changing the excitation wavelengths [85]. The CIE coordination of Eu^{3+}, Tb^{3+} singly- and co-doped fluorophosphate glasses are listed in Table 7 and also presented in Figure 11 as a function of the dopant concentration and excitation wavelength. The blue and green emissions of Tb^{3+} ions in addition to the red emission band of Eu^{3+} ions result in nearly white light generation at controlled doping and excitation energy. Therefore, Eu^{3+}-Tb^{3+} co-doped glasses are good candidates to provide tunable visible light when excited at UV frequencies. The white light emission can be provided thanks to the strong blue emission of Tb^{3+} ions at low concentration, a green emission from Tb^{3+} ions and a strong orange-red emission of Eu^{3+} ions which lie, respectively, around 436 nm (Tb^{3+}: $^5D_3 \rightarrow {}^7F_5$), 543 nm ($Tb^{3+}$: $^5D_4 \rightarrow {}^7F_5$) and 612 nm ($Eu^{3+}$: $^5D_0 \rightarrow {}^7F_2$). Fluorophosphate glasses containing various amount of the latter two rare earth ions having the same compositions as given in Table 4, by addition of REF_3 in expense of YF_3, where RE = Eu^{3+} or Tb^{3+} and x = 0, 0.1, 0.25, 0.5, 1.0, 2.0, 3.0 and 4.0 mol%; and co-doped samples with total x = 0.3, 0.5, 1.0 and 1.5 mol% (and equal amount of EuF_3 and TbF_3) were recently prepared by our group. The optical properties of singly- and co-doped samples were reported and the glasses show strong color tenability when excited in the 350-360 nm spectral region.

The singly Eu^{3+} doped fluorophosphate glasses show various emission bands when excited at 393 nm. Luminescence bands centered at around 577, 590, 612, 650 and 700 nm correspond to the radiative decays from $^5D_0$ excited state to ^7F_J (J = 0, 1, 2, 3, 4) lower lying state. The corresponding lifetime of the various concentrations of the Eu^{3+} ions are listed in Table 7, in addition to their calculated x and y CIE coordination. As can be seen, the excited state lifetime of the emission band centered at 612 nm is around 3.1-3.9 ms which is relatively higher than those phosphate and tellurite glasses [59], and their CIE coordination lies in the warm red region.

Emission spectra of the singly Tb^{3+} ions doped fluorophosphate glasses show various lines when excited at 375 nm, which are attributed to the transitions within the 4f shells of Tb^{3+} ions as 413 nm ($^5D_3 \rightarrow {}^7F_5$), 436 nm ($^5D_3 \rightarrow {}^7F_4$), 456 nm ($^5D_3 \rightarrow {}^7F_3$), 472 nm ($^5D_3 \rightarrow {}^7F_2$), 485 nm ($^5D_4 \rightarrow {}^7F_6$), and 543 nm ($^5D_4 \rightarrow {}^7F_5$). The intensity of the most important emission bands - correspond to the blue and green emission centered at 436 nm and 543 nm – depends strongly on the concentration of this rare earth ion. As discussed elsewhere [102], the energy transfer through the cross-relaxation mechanism between the energy states of Tb^{3+} ions increases by increasing the concentration of this dopant. Therefore, the blue-to-green emission intensity ratio decreases by increasing the concentration of Tb^{3+} ions. The broad excitation spectra of the Tb^{3+} ions and tenability of the blue-to-green emission intensity ratio assist to achieve tunable visible color when co-doped with strong red-emitting ions such as Eu^{3+} ions. Typical luminescence spectra of the EuF_3-TbF_3 co-doped fluorophosphate glass samples having 0.25mol% of each dopant, is shown in Figure 9. Such tunable excitation results in color tenability of the total emission in the visible region. Fig. 11 shows the CIE 1931 chromaticity diagram for various concentration of the Eu^{3+}, Tb^{3+} and Eu^{3+}-Tb^{3+} co-doped samples. As can be seen, the color of the spectrum could be tuned by either changing the concentration of the dopants or exciting at various wavelengths in the UV region.

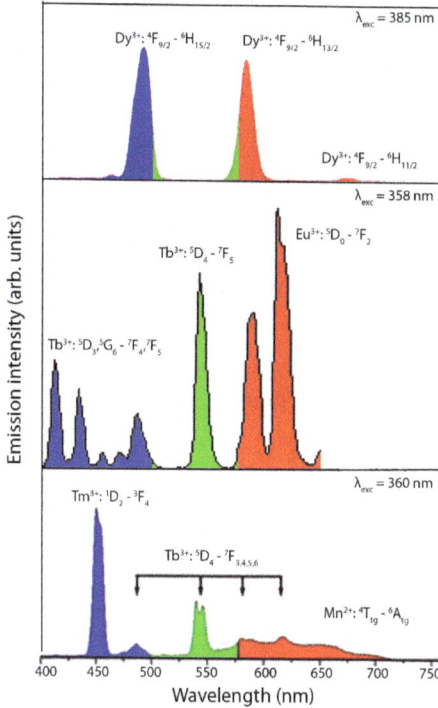

Figure 9. Emission spectra of Tb^{3+}/Mn^{2+}/Tm^{3+} tri-doped [101], Eu^{3+}-Tb^{3+} co-doped [85] and Dy^{3+} [27] singly-doped fluorophosphate glasses for white light generation. Excitation wavelengths are given in the legends.

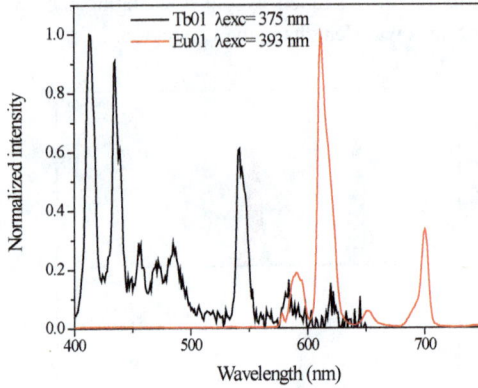

Figure 10. Emission spectra of the Eu^{3+} and Tb^{3+} doped fluorophosphate glasses doped with only 0.1 mol% of each lanthanide oxide.

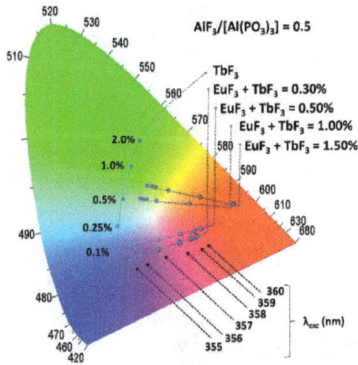

Figure 11. CIE chromaticity coordinates diagram showing the color variation as a function of excitation wavelength and dopant concentration. For samples singly doped with Tb^{3+} in the range 0.1 to 2.0 mol% excitation was performed at 355 nm. For co-doped samples with equal amounts of Eu^{3+} and Tb^{3+} and a total of 0.3, 0.5, 1.0 and 1.5 mol% REF_3 excitation was varied from 355 to 360 nm.

Table 7. Excited state lifetimes (τ) and chromaticity coordinate CIE(x, y) values for various Eu^{3+} or Tb^{3+} singly doped and co-doped samples. Data is taken from [85].

Excitation wavelength (nm)	EuF$_3$ (mol%)	TbF$_3$ (mol%)	$\tau(^5D_0)$ (ms)	$\tau(^5D_4)$ (ms)	x-CIE	y-CIE
393	0.1	-	3.1	-	0.648	0.327
393	0.25	-	3.3	-	0.634	0.327
393	0.5	-	3.3	-	0.645	0.329
393	1.0	-	3.4	-	0.654	0.327
393	2.0	-	3.5	-	0.671	0.327
393	3.0	-	3.7	-	0.654	0.329
393	4.0	-	3.9	-	0.673	0.325
376	-	0.1	-	3.7	0.247	0.239
376	-	0.25	-	3.7	0.250	0.299
376	-	0.5	-	3.7	0.265	0.370
376	-	1.0	-	3.7	0.289	0.458
376	-	2.0	-	3.8	0.313	0.527
376	-	3.0	-	4.0	0.314	0.545
376	-	4.0	-	4.5	0.312	0.525
355	0.15	0.15	-	-	0.285	0.222
	0.25	0.25	-	-	0.287	0.257
	0.50	0.50	-	-	0.315	0.380
	0.75	0.75	-	-	0.336	0.413
356	0.15	0.15	-	-	0.311	0.232
	0.25	0.25	-	-	0.311	0.261
	0.50	0.50	-	-	0.323	0.378
	0.75	0.75	-	-	0.349	0.411
357	0.15	0.15	-	-	0.363	0.249
	0.25	0.25	-	-	0.366	0.264
	0.50	0.50	-	-	0.334	0.377

	0.75	0.75	-	-	0.358	0.409
358	0.15	0.15	-	-	0.420	0.266
	0.25	0.25	-	-	0.425	0.288
	0.50	0.50	-	-	0.360	0.374
	0.75	0.75	-	-	0.393	0.401
359	0.15	0.15	-	-	0.452	0.277
	0.25	0.25	-	-	0.461	0.296
	0.50	0.50	-	-	0.449	0.367
	0.75	0.75	-	-	0.477	0.383
360	0.15	0.15	-	-	0.466	0.282
	0.25	0.25	-	-	0.480	0.301
	0.50	0.50	-	-	0.557	0.358
	0.75	0.75	-	-	0.566	0.366

Fluorophosphate glasses doped with one or more rare earth ions have been nominated as suitable candidates for white light generation. For example, Dy^{3+} doped $41P_2O_5$ - $17K_2O$ - $8 Al_2O_3$ - $24ZnF_2$ - $10LiF$ (mol%) glasses [103] show two strong emission bands at the yellow and the blue region which result in white light generation when excited at 386 nm. The x and y coordination of CIE-1931 diagram of such sample lies around x=0.31 and y=0.34 for 1 and 2 mol% of Dy_2O_3 doping. The lifetime of the $^4F_{9/2}$ excited state of Dy^{3+} ions are measured as listed in Table 8 in addition to the yellow to blue intensity ratio (Y/B). The exponential decay behavior of this level is only valid at low concentration, since at higher concentration, this state depopulates non-exponentially due to the cross-relaxation process between neighboring ions [27].

Table 8. Molar concentration (C), excited state lifetime of $^4F_{9/2}$ (τ), and Y/B intensity ratio of the Dy_2O_3 doped fluorophosphate glasses [27].

C (mol%)	τ (ms)	Y/B
0.01	0.750	0.56
0.05	0.778	0.57
0.1	0.779	0.59
0.5	0.715	0.61
1.0	0.572	0.63
2.0	0.374	0.61

4.   Summary

Fluorophosphate glasses have attracted great attention due to their excellent optical properties, chemical durability and thermal stability. They are good candidate as hosts for rare earth ions, provide good ion dispersivity for a large concentration of lanthanide oxides or fluorides. Therefore, in this chapter, the optical and structural properties of some new fluorophosphate glass composition were presented and appropriate references were added for an in-depth study of these glasses.

This chapter initially signified the importance of the glass science and technology, where the increase in the number of publication on glasses and particularly on fluorophosphate glasses indicates the fast research progress on the mentioned studies. Next, the advantage and disadvantages of both the phosphate based and fluorine based glasses are listed, and it is clarified that a suitable combination of these glass formers could result in relatively optimized fluorophosphate glass composition which benefit from the advantages of both the phosphate and fluorine network, while eliminating those detrimental parameters of both. Since the chapter deals with the spectroscopic properties of the lanthanide doped glasses, a brief introduction on the principals of ion-ion interactions and Judd-Ofelt analysis is given. Finally, a series of important results – published in recent years – on the structural and optical properties of lanthanide doped new fluorophosphate glasses for upconversion and near-infrared emissions as well as for the generation of white light and development of tunable solid state lightening materials are presented.

As discussed in this chapter, new fluorophosphate glasses were obtained with excellent optical and spectroscopic properties such as long lifetimes and high fluorescence quantum efficiencies. These glasses benefit from easy preparation, non-hygroscopicity, and wide transparency from the near infrared (4760 nm) to the UV (340 nm). Moreover, in the case of Er^{3+}-Yb^{3+} co-doped glasses, the energy transfer from Yb^{3+} to Er^{3+} ions are studied, and long lifetime and high quantum efficiency of broadband emission of Er^{3+} ions at around 1.53 μm are observed. Moreover, Tb^{3+}-Eu^{3+} co-doped glasses showed great potential as unable visible and/or white light generating materials, which can be selectively excited in the 355-360 nm UV spectral region. The excited state lifetime of both the ions is longer than previous reports on other glass compositions. At low concentration of dopants, blue and green emissions of Tb^{3+} ions and a red emission of Eu^{3+} ions results in the white light generation under UV excitation. Dy^{3+} singly-doped fluorophosphate glasses are also candidate for white light generation, which have a longer lifetime than other oxide glasses.

The Raman and NMR studies indicate that the improved optical properties of the new fluorophosphate glasses could be related to the dominating of the local coordination of the rare-earth ions by fluoride rather than phosphate species. Therefore, the exclusion of the OH species by fluorine improves the optical properties by decreasing the non-radiative decay rates.

Although, there are enormous numbers of publications on the lanthanide doped glasses, the structure/properties correlation in the rare earth ions doped fluorophosphate glasses deserves extra attention, due to the outstanding properties of them. Therefore, this chapter was dedicated to demonstrate the significance of such materials, and recent achievement on their optical and structural properties, concluding that a chemically durable, optically transparent material could be achieved for special application such as color tunable and white light generating solid state hosts.

5. Acknowledgments

The author would like to thank FAPESP for the post-doctoral fellowship process no. 13/24064-8 which is granted under the supervision of *Prof. Dr. Andrea Simone Stucchi de Camargo* (2014-2016). The works presented in this chapter has been taken from given references and data obtained in the Laboratory of Advanced Functional Materials (LEMAF, located in Instituto de Física de São Carlos, Universidade de São Paulo), and mainly achieved as the doctoral research project of *Tássia Gonçalves*, and collaboration of *Prof. Dr. Hellmut Eckert* which have been previously published as given in the bibliography of current chapter. Financial support from FAPESP as the CEPID project no. 13/07793-6, *CEPIV - Center for Teaching, Research and Innovation in Glass* is highly appreciated.

References

[1] E.D. Zanotto, A bright future for glass-ceramics, Am. Ceram. Soc. Bull. 89 (2010) 19–27.

[2] M.R. Dousti, R.J. Amjad, Plasmon assisted luminescence in rare earth doped glasses, in: C.D. Geddes (Ed.), Rev. Plasmon. 2015, Springer, 2016. http://dx.doi.org/10.1007/978-3-319-24606-2_14

[3] H. Tait, ed., Five thousand years of glass, The British Museum Press, London, 1991.

[4] M.J. Plodinec, Borosilicate glasses for nuclear waste imobilisation, Glas. Tech. 41 (2000) 186–192.

[5] L.B. Starling, J.E. Stephan, Calcium phosphate microcarries and microspheres, 6,358,532 B2, 2002.

[6] J.M. Clinton, W.W. Coffeen, Low Melting glasses in the system B_2O_3-ZnO-CaO-P_2O_5., Am. Ceram. Soc. Bull. 63 (1984) 1401–1404.

[7] J. Daimer, H. Paschke, Glass sealant containing lead borate glasss and fillers of ullite and cordierite, 5,145,803, 1990.

[8] L. Koudelka, P. Mošner, Study of the structure and properties of Pb–Zn borophosphate glasses, J. Non. Cryst. Solids. 293-295 (2001) 635–641. http://dx.doi.org/10.1016/S0022-3093(01)00765-7

[9] S.E. Stokowski, R. a Saroyan, M.J. Weber, Nd-Doped laser glass spectroscopic and physical properties, Livermore, CA, 1981.

[10] H.J. Stocker, Bulk and thin film switching and memory effects in semiconducting chalcogenide glasses, Appl. Phys. Lett. 15 (1969) 55. http://dx.doi.org/10.1063/1.1652900

[11] M. Poulain, Halide glasses, J. Non. Cryst. Solids. 56 (1983) 1–14. http://dx.doi.org/10.1016/0022-3093(83)90439-8

[12] C.M. Baldwin, R.M. Almeida, J.D. Mackenzie, Halide glasses, J. Non. Cryst. Solids. 43 (1981) 309–344. http://dx.doi.org/10.1016/0022-3093(81)90101-0

[13] Y. Hu, J. Qiu, Z. Song, D. Zhou, Ag_2O dependent up-conversion luminescence properties in $Tm^{3+}/Er^{3+}/Yb^{3+}$ co-doped oxyfluorogermanate glasses, J. Appl. Phys. 115 (2014) 083512. http://dx.doi.org/10.1063/1.4866875

[14] www.scopus.com, Elsevier B.V. (2014). www.scopus.com (accessed July 15, 2014).

[15] J.C. Mauro, E.D. Zanotto, Two centuries of glass research: historical trends, current status, and grand challenges for the future, Int. J. Appl. Glas. Sci. 15 (2014) 1–15. http://dx.doi.org/10.1111/ijag.12087

[16] G. Poirier, F.S. Ottoboni, F.C. Cassanjes, A. Remonte, Y. Messaddeq, S.J.L. Ribeiro, Redox behavior of molybdenum and tungsten in phosphate glasses., J. Phys. Chem. B. 112 (2008) 4481–7. http://dx.doi.org/10.1021/jp711709r

[17] R.K. Brow, Review : the structure of simple phosphate glasses, J. Non. Cryst. Solids. 263-264 (2000) 1–28. http://dx.doi.org/10.1016/S0022-3093(99)00620-1

[18] L.B. Fletcher, J.J. Witcher, N. Troy, S.T. Reis, R.K. Brow, R.M. Vazquez, et al., Femtosecond laser writing of waveguides in zinc phosphate glasses, Opt. Mater. Express. 1 (2011) 845. http://dx.doi.org/10.1364/OME.1.000845

[19] B. Peng, T. Izumitani, Blue, green and 0.8 μm Tm^{3+}, Ho^{3+} doped upconversion laser glasses, sensitized by Yb^{3+}, Opt. Mater. (Amst). 4 (1995) 701–711. http://dx.doi.org/10.1016/0925-3467(95)00031-3

[20] M. Nalin, S.J.L. Ribeiro, Y. Messaddeq, J. Schneider, P. Donoso, Scandium fluorophosphate glasses: a structural approach, Comptes Rendus Chim. 5 (2002) 915–920. http://dx.doi.org/10.1016/S1631-0748(02)01443-1

[21] D. Ehrt, Fluoroaluminate glasses for lasers and amplifiers, Curr. Opin. Solid State Mater. Sci. 7 (2003) 135–141. http://dx.doi.org/10.1016/S1359-0286(03)00049-4

[22] T.S. Gonçalves, R.J. Moreira Silva, M. de Oliveira Junior, C.R. Ferrari, G.Y. Poirier, H. Eckert, et al., Structure-property relations in new fluorophosphate glasses singly- and co-doped with Er^{3+} and Yb^{3+}, Mater. Chem. Phys. 157 (2015) 45–55. http://dx.doi.org/10.1016/j.matchemphys.2015.03.012

[23] M. Wang, L. Yi, Y. Chen, C. Yu, G. Wang, L. Hu, et al., Effect of $Al(PO_3)_3$ content on physical, chemical and optical properties of fluorophosphate glasses for 2μm application, Mater. Chem. Phys. 114 (2009) 295–299. http://dx.doi.org/10.1016/j.matchemphys.2008.09.014

[24] M.J. Dejneka, The luminescence and structure of novel transparent oxyfluoride glass-ceramics, J. Non. Cryst. Solids. 239 (1998) 149–155. http://dx.doi.org/10.1016/S0022-3093(98)00731-5

[25] P. A. Tick, N.F. Borrelli, L.K. Cornelius, M. a. Newhouse, Transparent glass ceramics for 1300 nm amplifier applications, J. Appl. Phys. 78 (1995) 6367–6374. http://dx.doi.org/10.1063/1.360518

[26] S.E. Stokowski, W.E. Martin, S.M. Yarema, Optical and lasing properties of fluorophosphate glass, J. Non. Cryst. Solids. 40 (1980) 481–487. http://dx.doi.org/10.1016/0022-3093(80)90123-4

[27] C.R. Kesavulu, C.K. Jayasankar, White light emission in Dy^{3+}-doped lead fluorophosphate glasses, Mater. Chem. Phys. 130 (2011) 1078–1085. http://dx.doi.org/10.1016/j.matchemphys.2011.08.037

[28] C.R. Kesavulu, K.K. Kumar, N. Vijaya, K.-S. Lim, C.K. Jayasankar, Thermal, vibrational and optical properties of Eu^{3+}-doped lead fluorophosphate glasses for red laser applications, Mater. Chem. Phys. 141 (2013) 903–911. http://dx.doi.org/10.1016/j.matchemphys.2013.06.021

[29] M. De Oliveira, T. Uesbeck, T.S. Gonçalves, C.J. Magon, P.S. Pizani, A.S.S. De Camargo, et al., Network Structure and Rare-Earth Ion Local Environments in Fluoride Phosphate Photonic Glasses Studied by Solid-State NMR and Electron Paramagnetic Resonance Spectroscopies, J. Phys. Chem. C. 119 (2015) 24574–24587. http://dx.doi.org/10.1021/acs.jpcc.5b08088

[30] R.K. Sandwick, R.J. Scheller, K.H. Mader, Production of high homogeneous fluorophosphate laser glass., Proc. Soc. Photo-Optical Instrum. Eng. 171 (1979) 161–116.

[31] J. Wysocki, M.J. Liepmann, Optical and mechanical properties of fluorophosphate glasses for, Proceeding SPIE. 1327 (1990) 238–249. http://dx.doi.org/10.1117/12.22539

[32] F. Gan, Y. Jiang, F. Jiang, Formation and structure of $Al(PO_3)_3$-containing fluorophosphate glass, J. Non. Cryst. Solids. 52 (1982) 263–273. http://dx.doi.org/10.1016/0022-3093(82)90301-5

[33] B. Karmakar, K. Annapurna, Blue, green and red upconversions in Ho_2O_3-doped fluorophosphate glasses, J. Non. Cryst. Solids. 353 (2007) 1377–1382. http://dx.doi.org/10.1016/j.jnoncrysol.2006.09.057

[34] S. Stevic, R. Aleksic, N. Backovi, Influence of Fluorine on Thermal Properties of Fluorophosphate Glasses, J. Am. Ceram. Soc. 70 (1987) 264–265. http://dx.doi.org/10.1111/j.1151-2916.1987.tb04894.x

[35] S. Liu, A. Lu, Physical and spectroscopic properties of Yb^{3+}-doped fluorophosphate laser glasses, Laser Chem. 2008 (2008) 1–6. http://dx.doi.org/10.1155/2008/656490

[36] M. Liao, Z. Duan, L. Hu, Y. Fang, L. Wen, Spectroscopic properties of Er^{3+}/Yb^{3+} codoped fluorophosphate glasses, J. Lumin. 126 (2007) 139–144. http://dx.doi.org/10.1016/j.jlumin.2006.06.009

[37] M. Liao, H. Sun, L. Wen, Y. Fang, L. Hu, Effect of alkali and alkaline earth fluoride introduction on thermal stability and structure of fluorophosphate glasses, Mater. Chem. Phys. 98 (2006) 154–158. http://dx.doi.org/10.1016/j.matchemphys.2005.09.006

[38] N. Rigout, J.L. Adam, J. Lucas, Chemical and physical compatibilities of fluoride and fluorophosphate glasses, J. Non. Cryst. Solids. 184 (1995) 319–323. http://dx.doi.org/10.1016/0022-3093(94)00590-7

[39] T. Izumitani, Y. Asahara, Cause of low damage threshold of fluorophosphate glass, AGARD Lect. Ser. (1980) 172–179.

[40] K. Binnemans, R. Van Deun, C. Görller-Walrand, J.L. Adam, Optical properties of Nd^{3+}-doped fluorophosphate glasses, J. Alloys Compd. 275-277 (1998) 455–460. http://dx.doi.org/10.1016/S0925-8388(98)00367-3

[41] C.W. Thiel, Y. Sun, R.L. Cone, Photonic materials and devices progress in relating rare-earth ion 4f and 5d energy levels to host bands in optical materials for hole burning, quantum information and phosphors, J. Mod. Opt. 49 (2002) 2399–2411. http://dx.doi.org/10.1080/0950034021000011491

[42] A. Jha, B. Richards, G. Jose, T. Teddy-Fernandez, P. Joshi, X. Jiang, et al., Rare-earth ion doped TeO_2 and GeO_2 glasses as laser materials, Prog. Mater. Sci. 57 (2012) 1426–1491. http://dx.doi.org/10.1016/j.pmatsci.2012.04.003

[43] P.C. Becker, N.A. Olsson, J.R. Simpson, Erbium-doped fiber amplifiers, Academic Press, San Diego, 1999.

[44] K. Maheshvaran, K. Marimuthu, Concentration dependent Eu^{3+} doped boro-tellurite glasses-Structural and optical investigations, J. Lumin. 132 (2012) 2259–2267. http://dx.doi.org/10.1016/j.jlumin.2012.04.022

[45] S. Dai, C. Yu, G. Zhou, J. Zhang, G. Wang, L. Hu, Concentration quenching in erbium-doped tellurite glasses, J. Lumin. 117 (2006) 39–45. http://dx.doi.org/10.1016/j.jlumin.2005.04.003

[46] H. Zheng, D. Gao, Z. Fu, E. Wang, Y. Lei, Y. Tuan, et al., Fluorescence enhancement of Ln^{3+} doped nanoparticles, J. Lumin. 131 (2011) 423–428. http://dx.doi.org/10.1016/j.jlumin.2010.09.026

[47]   E. Snitzer, Optical maser action of Nd^{3+} in a barium crown glass, Phys. Rev. Lett. 7 (1961).

[48]   C.J. Koester, E. Snitzer, Amplification in a Fiber Laser, Appl. Opt. 3 (1964) 1182. http://dx.doi.org/10.1364/AO.3.001182

[49]   K.C. Kao, Dielectric-fibre surface waveguides for optical frequencies, Proceeding Inst Electr Eng-London. 113 (1966) 1151. http://dx.doi.org/10.1049/piee.1966.0189

[50]   J.N. Sandoe, P.H. Sarkies, S. Parke, Variation of Er^{3+} cross-section for stimulated emission with glass composition., J. Phys. D. Appl. Phys. 5 (1972) 1788. http://dx.doi.org/10.1088/0022-3727/5/10/307

[51]   J. Stone, C.A. Burrus, Neodymium-doped silica lasers in end-pumped fiber geometry., Appl. Phys. Lett. 23 (1973) 388–9. http://dx.doi.org/10.1063/1.1654929

[52]   R.J. Mears, L. Reekie, S.B. Poole, D.N. Payne, Low-threshold tunable CW and Q-switched fibre laser operating at 1.55µm, Electron. Lett. 22 (1986) 159–160. http://dx.doi.org/10.1049/el:19860111

[53]   F. Auzel, Upconversion processes in coupled ion systems, J. Lumin. 45 (1990) 341–345. http://dx.doi.org/10.1016/0022-2313(90)90189-I

[54]   D.L. Dexter, A Theory of Sensitized Luminescence in Solids, J. Chem. Phys. 21 (1953) 836–851. http://dx.doi.org/10.1063/1.1699044

[55]   A.A. Kaplyanskii, R.M. MacFarlane, eds., Spectroscopy of Solids containing Rare-Earth Ion, North-Holland, Amsterdam, 1987.

[56]   B.R. Judd, Optical absorption intensities of rare-earth ions, Phys. Rev. 127 (1962) 750–761. http://dx.doi.org/10.1103/PhysRev.127.750

[57]   G.S. Ofelt, Intensities of crystal spectra of rare-earth ions, J. Chem. Phys. 37 (1962) 511–520. http://dx.doi.org/10.1063/1.1701366

[58]   B.M. Walsh, Judd-ofelt theory: principles and practices, in: Adv. Spectrocopy Lasers Sens., 2006: pp. 403–433.

[59]   M.R. Dousti, G.Y. Poirier, A.S.S. de Camargo, Structural and spectroscopic characteristics of Eu^{3+}-doped tungsten phosphate glasses, Opt. Mater. (Amst). 45 (2015) 185–190. http://dx.doi.org/10.1016/j.optmat.2015.03.033

[60]   X. Li, B. Chen, R. Shen, H. Zhong, L. Cheng, J. Sun, et al., Fluorescence quenching of ^5D_J (J=1, 2 and 3) levels and Judd–Ofelt analysis of Eu^{3+} in

NaGdTiO$_4$ phosphors, J. Phys. D. Appl. Phys. 44 (2011) 335403. http://dx.doi.org/10.1088/0022-3727/44/33/335403

[61] H.U. Rahman, Optical intensities of trivalent erbium in various host lattices, J. Phys. C Solid State Phys. 5 (1972) 306–315. http://dx.doi.org/10.1088/0022-3719/5/3/010

[62] W.T. Carnall, H. Crosswhite, H.M. Crosswhite, Energy level structure and transition probabilities in the spectra of the trivalent lanthandes in LaF$_3$, 1977.

[63] J. Yang, S. Dai, N. Dai, L. Wen, L. Hu, Z. Jiang, Investigation on nonradiative decay of $^4I_{13/2}$ - $^4I_{15/2}$ transition of Er^{3+} -doped oxide glasses, J. Lumin. 106 (2004) 9–14. http://dx.doi.org/10.1016/S0022-2313(03)00128-5

[64] Y. Chen, Y. Huang, M. Huang, R. Chen, Z. Luo, Spectroscopic properties of Er^{3+} ions in bismuth borate glasses, Opt. Mater. (Amst). 25 (2004) 271–278. http://dx.doi.org/10.1016/j.optmat.2003.07.002

[65] P. Nandi, G. Jose, Spectroscopic properties of Er^{3+} doped phospho-tellurite glasses, Phys. B Condens. Matter. 381 (2006) 66–72. http://dx.doi.org/10.1016/j.physb.2005.12.255

[66] M.D. Shinn, W. a. Sibley, M.G. Drexhage, R.N. Brown, Optical transitions of Er^{3+} ions in fluorozirconate glass, Phys. Rev. B. 27 (1983) 6635–6648. http://dx.doi.org/10.1103/PhysRevB.27.6635

[67] D.K. Sardar, J.B. Gruber, B. Zandi, J.A. Hutchinson, C. Ward Trussell, Judd-Ofelt analysis of the Er^{3+}(4f11) absorption intensities in phosphate glass: Er^{3+}, Yb^{3+}, J. Appl. Phys. 93 (2003) 2041–2046. http://dx.doi.org/10.1063/1.1536738

[68] R. Van Deun, K. Binnemans, C. Görller-Walrand, J.L. Adam, Judd–Ofelt intensity parameters of trivalent lanthanide ions in a NaPO$_3$–BaF$_2$ based fluorophosphate glass, J. Alloys Compd. 283 (1999) 59–65. http://dx.doi.org/10.1016/S0925-8388(98)00895-0

[69] E.L. Raaben, A.K. Przhevuskii, S.G. Lunter, Probabilities of the optical transitions of neodymium in a fluorophosphate glass, J. Appl. Spectrosc. 24 (1976) 179–183. http://dx.doi.org/10.1007/BF00612273

[70] O. Deutschbein, M. Faulstich, W. Jahn, G. Krolla, N. Neuroth, Glasses with a large laser effect : Nd-phosphate and, Appl. Opt. 17 (1978) 2228–2232. http://dx.doi.org/10.1364/AO.17.002228

[71] S. Buddhudu, Characterization of Fluorophosphate Optical Glasses, 460 (1991) 454–460.

[72] J. Chrysochoos, B. Kumar, S.P. Sinha, Time-resolved luminescence and decay characteristics of Gd^{3+} in fluoroarsenate and fluorophosphate glasses, J. Less Common Met. 126 (1986) 195–201. http://dx.doi.org/10.1016/0022-5088(86)90280-8

[73] M. Sreenivasulu, A.S. Rao, Absorption and emission spectra of Pr^{3+}-doped mixed alkali fluorophosphate optical glasses, (2001) 737–740.

[74] S. Tanabe, K. Hirao, N. Soga, Local structure of rare-earth ions in fluorophosphate glasses by phonon sideband and mössbauer spectroscopy, J. Non. Cryst. Solids. 142 (1992) 148–154. http://dx.doi.org/10.1016/S0022-3093(05)80017-1

[75] R.P.R.D. Nardi, C.E. Braz, A.S.S. de Camargo, S.J.L. Ribeiro, L.A. Rocha, F.C. Cassanjes, et al., Effect of lead fluoride incorporation on the structure and luminescence properties of tungsten sodium phosphate glasses, Opt. Mater. 49 (2015) 249–254. http://dx.doi.org/10.1016/j.optmat.2015.09.008

[76] A.V.R. Reddy, T. Balaji, S. Buddhudu, Absorption and photoluminescence spectra of Tm^{3+}-doped fluorophosphate glasses, 1992.

[77] G. Özen, J.P. Denis, P. Goldner, X. Wu, M. Genotelle, F. Pellé, Enhanced Tm^{3+} blue emission in Tm, Yb, co-doped fluorophosphate glasses due to back energy transfer processes, Appl. Phys. Lett. 62 (1993) 928. http://dx.doi.org/10.1063/1.108522

[78] H. Ono, K. Nakagawa, M. Yamada, S. Sudo, Er^{3+}-doped fluorophosphate glass fibre amplifier for WDM systems, Electron. Lett. 32 (1996) 1586–1587. http://dx.doi.org/10.1049/el:19961038

[79] Z. Li-Yan, H. Li-Li, Evaluation of broadband spectral properties of erbium-doped aluminium fluorophosphate glass, Chinese Phys. Lett. 20 (2003) 1836–1837. http://dx.doi.org/10.1088/0256-307X/20/10/351

[80] H. Sun, L. Zhang, S. Xu, S. Dai, J. Zhang, L. Hu, et al., Structure and thermal stability of novel fluorophosphate glasses, J. Alloys Compd. 391 (2005) 151–155. http://dx.doi.org/10.1016/j.jallcom.2004.07.071

[81] R. Zheng, Z. Wang, P. Lv, Y. Yuan, Y. Zhang, J. Zheng, et al., Novel synthesis of low hydroxyl content Yb^{3+}-doped fluorophosphate glasses with long fluorescence lifetimes, J. Am. Ceram. Soc. 98 (2015) 861–866. http://dx.doi.org/10.1111/jace.13386

[82] L. Zhang, Y. Leng, J. Zhang, L. Hu, Yb^{3+}-doped fluorophosphate glass with high cross section and lifetime, J. Mater. Sci. Technol. 26 (2010) 921–924. http://dx.doi.org/10.1016/S1005-0302(10)60148-X

[83] J. Qiu, K. Tanaka, N. Sugimoto, K. Hirao, Faraday effect in Tb^{3+}-containing borate, fluoride and fluorophosphate glasses, J. Non. Cryst. Solids. 213-214 (1997) 193–198. http://dx.doi.org/10.1016/S0022-3093(97)00101-4

[84] T. Asahara, YoshiyukiIzumitani, Faraday Rotation Glass, 1983.

[85] T.B. De Queiroz, M.B.S. Botelho, T.S. Gonçalves, M.R. Dousti, A.S.S. De Camargo, New fluorophosphate glasses co-doped with Eu^{3+} and Tb^{3+} as candidates for generating tunable visible light, J. Alloys Compd. 647 (2015) 315–321. http://dx.doi.org/10.1016/j.jallcom.2015.06.066

[86] B. Karmakar, P. Kundu, R. Dwivedi, IR spectra and their application for evaluating physical properties of fluorophosphate glasses, J. Non. Cryst. Solids. 289 (2001) 155–162. http://dx.doi.org/10.1016/S0022-3093(01)00721-9

[87] G.J. Exarhos, P.J. Miller, W.M. Risen, Interionic vibrations and glass transitions in ionic oxide metaphosphate glasses, J. Chem. Phys. 60 (1974) 4145–4155. http://dx.doi.org/10.1063/1.1680881

[88] D.E.C. Corbridge, E.J. Lowe, The infra-red spectra of some inorganic phosphorus compounds, L. Chem. Soc. (1954) 493–502.

[89] K. Nakamoto, Infrared and Raman spectra of inorganic and coordination compounds, 4th ed., Wiley, New York, 1986.

[90] R.K. Brow, Z.A. Osborne, R.J. Kirkpatrick, Multinuclear MAS NMR study of the short-range structure of fluorophosphate glass, J. Mater. Res. 7 (1992) 1892–1899. http://dx.doi.org/10.1557/JMR.1992.1892

[91] T. Djouama, M. Poulain, B. Bureau, R. Lebullenger, Structural investigation of fluorophosphate glasses by ^{19}F, ^{31}P MAS-NMR and IR spectroscopy, J. Non. Cryst. Solids. 414 (2015) 16–20. http://dx.doi.org/10.1016/j.jnoncrysol.2015.01.017

[92] B. Karmakar, P. Kundu, R.N. Dwivedi, UV transparency and structure of fluorophosphate glasses, Mater. Lett. 57 (2002) 953–958. http://dx.doi.org/10.1016/S0167-577X(02)00903-5

[93] J.H. Choi, F.G. Shi, A. Margaryan, Refractive index and low dispersion Properties of new fluorophosphate glasses highly doped with rare-earth ions, J. Mater. Res. 20 (2011) 264–270. http://dx.doi.org/10.1557/JMR.2005.0033

[94]   G. Poirier, C.B. De Araújo, Y. Messaddeq, S.J.L. Ribeiro, M. Poulain, Tungstate fluorophosphate glasses as optical limiters, J. Appl. Phys. 91 (2002) 10221–10223. http://dx.doi.org/10.1063/1.1481770

[95]   G.G. Poirier, V. a. Jerez, C.B. de Araújo, Y. Messaddeq, S.J.L. Ribeiro, M. Poulain, et al., Optical spectroscopy and frequency upconversion properties of Tm^{3+} doped tungstate fluorophosphate glasses, J. Appl. Phys. 93 (2003) 1493. http://dx.doi.org/10.1063/1.1536017

[96]   J.H. Choi, F.G. Shi, A. Margaryan, A. Margaryan, W. van der Veer, Novel alkaline-free Er^{3+}-doped fluorophosphate glasses for broadband optical fiber lasers and amplifiers, J. Alloys Compd. 450 (2008) 540–545. http://dx.doi.org/10.1016/j.jallcom.2007.07.094

[97]   S.S. Babu, P. Babu, C.K. Jayasankar, T. Tröster, W. Sievers, G. Wortmann, Optical properties of Dy^{3+}-doped phosphate and fluorophosphate glasses, Opt. Mater. 31 (2009) 624–631. http://dx.doi.org/10.1016/j.optmat.2008.06.019

[98]   M.R. Dousti, Origins of the broadening in 1.5 μm emission of Er^{3+}-doped glasses, J. Mol. Struct. 1100 (2015) 415–420. http://dx.doi.org/10.1016/j.molstruc.2015.07.063

[99]   N. Jaba, H. Ben Mansour, A. Kanoun, A. Brenier, B. Champagnon, Spectral broadening and luminescence quenching of 1.53μm emission in Er^{3+}-doped zinc tellurite glass, J. Lumin. 129 (2009) 270–276. http://dx.doi.org/10.1016/j.jlumin.2008.10.006

[100] R. Deng, F. Qin, R. Chen, W. Huang, M. Hong, X. Liu, Temporal full-colour tuning through non-steady-state upconversion, Nat Nanotechnol. 10 (2015) 237–242. http://dx.doi.org/10.1038/nnano.2014.317

[101] C. Ming, F. Song, Y. Qin, X. Ren, L. An, M (Tm^{3+}, Tb^{3+}, Ho^{3+}, Dy^{3+}, Mn^{2+})-doped transparent fluorophosphate glasses for white light-emitting-diodes, Opt. Commun. 321 (2014) 195–197. http://dx.doi.org/10.1016/j.optcom.2014.02.013

[102] J.F.M. dos Santos, I. a. a. Terra, N.G.C. Astrath, F.B. Guimarães, M.L. Baesso, L. a. O. Nunes, et al., Mechanisms of optical losses in the $^5D_4$ and $^5D_3$ levels in Tb^{3+} doped low silica calcium aluminosilicate glasses, J. Appl. Phys. 117 (2015) 053102. http://dx.doi.org/10.1063/1.4906781

[103] N. Vijaya, K. Upendra Kumar, C.K. Jayasankar, Dy^{3+}-doped zinc fluorophosphate glasses for white luminescence applications, Spectrochim. Acta - Part A Mol.

Biomol. Spectrosc. 113 (2013) 145–153.
http://dx.doi.org/10.1016/j.saa.2013.04.036

CHAPTER 2

Lanthanides co-doped phosphate glasses for broadband applications

Hssen Fares[1], Habib Elhouichet[1, 2]

[1]Laboratoire de Physico-Chimie des Matériaux Minéraux et leurs Applications, Centre National de Recherches en Sciences des Matériaux, B.P. 95 Hammam-Lif, 2050, Tunisia

[2]Département de Physique, Faculté des Sciences de Tunis, University Tunis ElManar 2092, Tunisia

Abstract

Erbium (Er^{3+}) doped phosphate glasses are striking materials due to their broadband emission at around 1.55 μm, thanks to various transitions from $^4I_{13/2}$ state to $^4I_{15/2}$ manifolds. Few characteristics, such as effective bandwidth ($\Delta\lambda_{eff}$), full-width at half-maximum (FWHM), emission cross-section (σ_e), and bandwidth quality factor (FWHM×σ_e) are determinant parameters to obtain practical appropriate materials for broadband applications such as Erbium-doped fiber amplifiers (EDFAs). In this chapter, we first revisit the physical origins for the broadening of the 1.55 μm emission. The effect of other rare earth ions RE in the 1.55 μm emission of the Er^{3+} doped phosphate glasses is also investigated. More important, incorporation of metallic nanoparticles in Er^{3+} doped phosphate glasses has been introduced as an interesting method to enhance their optical properties for broadband applications.

Keywords

Phosphate Glasses, Erbium Ions, Lanthanides Dependent Broadening, Broadband Applications, Metallic Nanoparticles Effect

Contents

1.   Introduction

The necessity to develop new materials for various applications such as optoelectronic devices, military facilities, optical devices and medical interferences has motivated the materials engineers and scientists to demand for noble materials, with low-cost, high efficiency, long durability and recyclability. Glass technology is classified to be among the most important branch in the materials science since these composites could provide a wide range of optical, thermal, chemical and electrical properties due to the large diversity in their structure and compositional elements. In fact, glasses have emerged in various fields and show significant contribution in the development of new functional materials. It can be used as laser hosts, switching panels, memory planes, optical fibers, waveguides, sensors, optical limiters, army devices and medical devices etc. The research on the area of glasses shows a rapid growth as evaluated by a recent analysis of the major scientific databases. The glasses are widely used as optical materials as they usually show a wide transparency from the ultraviolet (UV) to near-infrared (NIR) region and non-linear optical absorption, which could be combined with their other promising properties to obtain novel materials like rare earth and metallic particles (such as nanoparticles, clusters, atoms, dimers, ions, etc). In fact, Glasses doped with rare earth (RE) ions have been extensively investigated due to their laser action in visible and near infrared regions. The signal generated within the RE emission band stimulates emission of radiation at the same frequency, amplifying the optical communication signal with high gain, high efficiency and low noise, which is highly advantageous for optical communications [1]. There are five main RE candidates for use as dopants in fiber or waveguide amplifiers for optical communications systems: Er^{3+}, $Tm^{3+,}$ Nd^{3+}, Pr^{3+} and Dy^{3+}. The Er^{3+} and Tm^{3+} ions are the choice for the 1400-1600 nm window centered at 1550 nm, based on the $^4I_{13/2} \rightarrow {}^4I_{15/2}$ transition of Er^{3+} ion and the $^3H_4 \rightarrow {}^3F_4$ transition of Tm^{3+} ion. The $^4F_{3/2} \rightarrow {}^4I_{13/2}$ emission of Nd^{3+} ion, the $^1G_4 \rightarrow {}^3H_5$ transition of Pr^{3+} ion and the $^6F_{11/2} \rightarrow {}^6H_{15/2}$ transition of Dy^{3+} ion are all potentially useful for the 1300 nm telecommunication window. Special attention has been given to trivalent erbium (Er^{3+}) and in particular its emission band around 1550 nm. This emission is within the telecommunication windows, and fortuitously coincides with the 1550 nm intra-4f $^4I_{13/2} \rightarrow ^4I_{15/2}$ transition of the Er^{3+} ion, the reason behind the existence of the great interest in utilizing erbium-doped materials for gain elements and sources in telecommunications systems today [2, 3]. The development of the erbium-doped fiber amplifier (EDFA) in the late 1980s [4, 5] exploited the $^4I_{13/2} \rightarrow ^4I_{15/2}$ transition and allowed the transmission and amplification of signals in the 1530-1560 nm region without the necessity for expensive optical to electrical conversion [6]. Hence, it became a driving force for research in RE doped fibers and integrated optics waveguides have been used for amplifying weak signals in optical communications systems at 1300 and 1600 nm. A lot of work has been carried out over the past several years on the spectral analysis of Er^{3+} ions in a variety of glassy matrices including borate [7], tellurite [8-10], bismuth [11] and phosphate [12, 13]. In general, application and utilities of glassy materials are enormous and are governed by several factors including thermal stability, phonon energy, refractive index and dopants/ impurities present in the matrix. Moreover, the intensity, peak wavelength, excited state lifetime and quantum efficiency of these emissions are determinant characteristics of the RE-doped materials which could be engineered by selection of the host glass, concentration of RE ions, the co-doping species, heat-treatments etc. For example, the selection of the glass hosts with low phonon energies, can reduce the non-radiative losses and increases the quantum efficiency, while up-conversion emissions could occur to generate a high energy photon through the excitation of the RE ions by two incoming photons. Among different glass hosts such as silicate, tellurite, borate, etc…, phosphate glasses have their unique characteristics that include high transparency, low melting point, high thermal stability, high gain density, due to high solubility for lanthanide (Ln) ions, low refractive index, low glass transition temperature, low optical dispersion and high thermal expansion coefficients [14, 15]. These properties make it a suitable laser host and are essential for the fabrication of compact waveguide laser devices and other vital applications such as glass-to metal seals and bio-compatible materials [16]. In fact, phosphate glass shows great potential and can be used as a bio-material due to the fact that its chemical composition can be easily adjusted to be close to that of natural bone. However, one drawback of this glass is that it is hygroscopic and the OH^- group, which acts like high-energy phonons, can severely affect the quantum efficiency of the emitting rare-earth ions. The incorporation of an OH^- group is mainly due to the starting material and the atmospheric moisture during melting, hence there are ways to reduce its incorporation in the glass by composition variation and during the preparation process [16]. Our recent extensive studies on Er^{3+} doped phosphate glasses reveal the potential of these matrixes for an efficient 1.55 µm broadband optical amplifiers [12, 13].

Here, we primarily discuss the origins of the broadening in $^4I_{13/2} \rightarrow ^4I_{15/2}$ transition of Er^{3+} doped glasses.

2. Origins of the broadening in $^4I_{13/2} \rightarrow ^4I_{15/2}$ transition of Er^{3+} doped glasses

To provide a comprehensive evaluation of the emission properties of the Er^{3+} ions doped phosphate glasses, many researchers have estimated important parameters, those which are relevant for optical amplifiers for $^4I_{13/2} \rightarrow ^4I_{15/2}$ transition such as effective bandwidth ($\Delta\lambda_{eff}$), full-width at half-maximum (FWHM), emission cross-section (σ_e), and bandwidth quality factor (FWHM×σ_e) which are important parameters for the Er^{3+}-doped fiber amplifier (EDFA) used in the wavelength division multiplexing (WDM) network system of optical communication.

The effective bandwidth ($\Delta\lambda_{eff}$), can be expressed according to Weber et al by [17]:

$$\Delta\lambda_{eff} = \int \frac{I(\lambda)d\lambda}{I_{max}} \tag{1}$$

where $I(\lambda)$ is the emission intensity at the wavelength λ, and I_{max} is the intensity at the peak wavelength λ.

The stimulated emission cross-section (σ_e) of Er^{3+} and the absorption cross section (σ_a) are related by the Mc-Cumber theory [18] expressed as follow:

$$\sigma_e(\upsilon) = \sigma_a(\upsilon)\exp\left(\frac{(\varepsilon - h\upsilon)}{KT}\right) \tag{2}$$

where σ_a is the absorption cross-section, υ is the photon frequency, h the Planck constant, k the Boltzmann constant, and ε the net free energy required to excite one Er^{3+} ion from the $^4I_{15/2}$ to $^4I_{13/2}$ state at temperature T.

Erbium ions can be excited at several wavelengths including 488 and 980 nm. The NIR luminescence mechanisms for the Er^{3+}doped samples could be explained on the basis of the energy level diagram of Er^{3+} ions, which is shown in Fig.1.

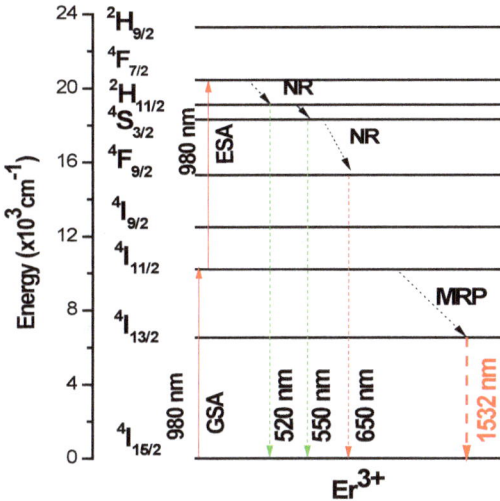

Figure 1: Schematic energy level diagram of Er³⁺ ion and luminescence mechanisms under the 980 nm excitation.

First, a 980 nm laser beam excites the 4f electrons of Er^{3+} ions at the ground state to the excited state $^4I_{11/2}$ by ground state absorption (GSA). Then, they can be excited to $^4F_{7/2}$ level by absorption of another photon via excited state absorption (ESA). Subsequently, the electrons populated rapidly from $^4F_{7/2}$ level to $^2H_{11/2}$, $^4S_{3/2}$ and $^4F_{9/2}$ levels by non-radiative (NR) decays, which produces green emissions at 520 nm ($^2H_{11/2} \rightarrow ^4I_{15/2}$) and 550 nm ($^4S_{3/2} \rightarrow ^4I_{15/2}$) followed by red emission at 650 nm due to the transition $^4F_{9/2} \rightarrow ^4I_{15/2}$ [9]. During GSA, the electrons at $^4I_{11/2}$ level of Er^{3+} can also relax to $^4I_{13/2}$ level by non-radiative multi-phonon relaxation (MRP), which leads to three possible emissions: (1) producing 1532 nm emission by the transition $^4I_{13/2} \rightarrow ^4I_{15/2}$, (2) producing 650 nm emission by absorbing another photon from adjacent Er^{3+} ions $^4I_{13/2} \rightarrow ^4F_{9/2} \rightarrow ^4I_{15/2}$ (two-photon process) and (3) producing 408 nm emission by absorbing two other photons by the transition $^4I_{13/2} \rightarrow ^4F_{9/2} \rightarrow ^2H_{9/2} \rightarrow ^4I_{15/2}$ (three-photon process) [9].

In general, the broadening of the 1.53 μm emission of the Er^{3+} doped glasses is due to the effect of the spin-orbit coupling and crystal field on the energy splitting of the ^4I isolated energy level of Er^{3+} ions. In fact, depopulation of $^4I_{13/2}$ excited state of Er^{3+} ions to its ground state ($^4I_{15/2}$) generates a broadband emission due to the large Stark splitting of these two manifolds in the crystal field of the surrounding medium. The average energy difference between adjacent Stark levels is around 50 cm^{-1} [19], resulting in an overall spread of 300-400 cm^{-1} within each state. Fig. 2 shows the effect of the spin-orbit coupling and crystal field on the energy splitting of the ^4I isolated energy level of Er^{3+} ions.

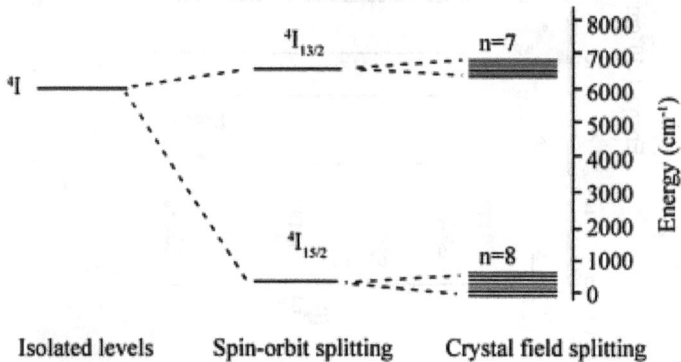

Figure 2: The effect of spin-orbit and crystal field splitting on the energy levels of Er^{3+} ions in silicate glass. Figure is adopted from Ref. [20].

It is worth mentioning that the broadening of the latter near-infrared emission of erbium ions is also related to the excitation energy. According to Coelho et al. [21], the peak position of NIR emission shifts from 1532 to 1540 nm by varying the excitation wavelength from 970 to 495 nm, whereas the bandwidth increases from 67 to 85 nm, respectively. Such behavior was related to the presence of non-radiative channels through ion-phonon interactions. However, Fig. 3 shows the NIR band of Er^{3+} doped tungsten phosphate glass at different excitation wavelengths. As we can see, the emission band centered at 1532 nm is larger when excited at 980 nm than that 486 nm. According to M. Reza Dousti [22], such excitation energy dependency of the shape of this emission might be associated to the selective excitation of the active ions and non-radiative processes such as cross-relaxations. These observations are not fully understood yet, and need rethinking and re-evaluation in order to ascertain its origins.

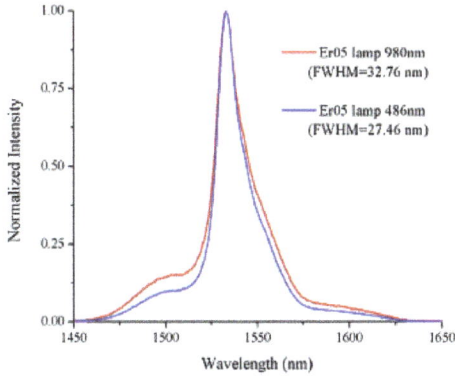

Figure 3: NIR emission of 0.5mol% Er^{3+}-doped tungsten-phosphate glass under different excitation wavelengths. Figure is adopted from Ref. [22]

The broadening of the near-infrared emission of erbium ions, lifetime and FWHM also depend on the type and amount of modifiers, which strongly affects the shape of the emission spectrum due to the character of the ion-host interactions. Recently, the emission properties of $^4I_{13/2} \rightarrow {}^4I_{15/2}$ transition in phosphate glasses doped with Er^{3+} ions have been examined as a function of PbF$_2$ concentration [23]. It is seen from Fig. 4 (a) that the luminescence linewidth, defined as the half maximum (FWHM), increases with increasing PbF$_2$ concentration in phosphate glass from 50 nm to 109 nm. In general, glass material is a disorder medium offering a broad diversity of sites, the site-to-site variation of Er^{3+} ions is also a main factor to affect the bandwidth. Thus, the observed linewidths are a convolution of several inhomogeneously broadened transitions between individual crystal-field levels and different Er^{3+} ions located in different sites. According to authors, the Er^{3+} ions incorporate with the interstitial F- ions located in the neighboring sites, resulting in a variation of the crystal field at the Er^{3+} sites which contributes to the inhomogeneous broadening of the luminescence band at 1530 nm [23].

51

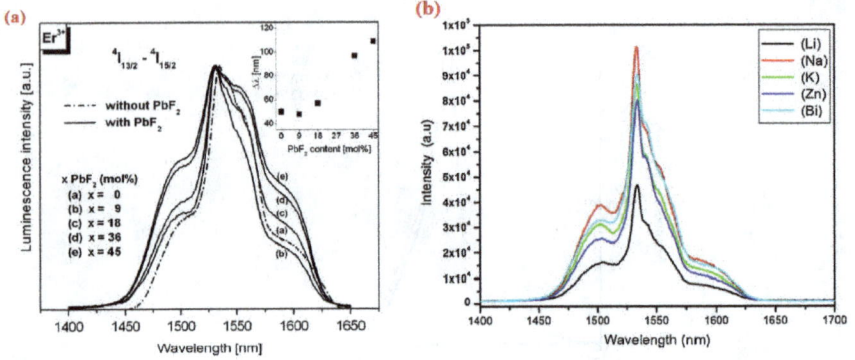

Figure 4: (a) NIR emission spectra for Er $^{3+}$ doped phosphate glasses modified by PbF2. (b) NIR emission spectra of Er^{3+} doped different fluoro-phosphate glasses. Figures is adopted respectively from Refs [23, 24].

In another study, optical properties of Er^{3+} doped fluoro-phosphate (FP) glasses with various additives such as alkali oxides (Li$_2$O, Na$_2$O, and K$_2$O), metal oxides (ZnO) and heavy metal oxides (Bi$_2$O$_3$) is reported by S. Babu [24]. The intension of the present work is to examine the broadening of $^4I_{13/2} \rightarrow ^4I_{15/2}$ transition in phosphate glasses by changing the modifiers. The results shown in Fig. 4 (b) indicate that modifiers ions have a strong influence in modulating the intensity of peak as well as in the broadening of the near-infrared emission of Er^{3+} doped fluoro-phosphate glass samples. In fact, in the lithium FP glass matrix, emission bandwidth ($\Delta\lambda_{eff}$) for the $^4I_{13/2} \rightarrow ^4I_{15/2}$ transition is higher than that of other FP glasses which suggests that inhomogeneous broadening is high in lithium FP glass matrix. On the other hand, a low inhomogeneous broadening is observed in the bismuth FP glass matrix. They concluded this by changing the addition of glass modifier ions in P$_2$O$_5$-AlF$_3$-BaF$_2$-SrF$_2$-PbO-Er$_2$O$_3$-M (M = Li$_2$O, Na$_2$O, K$_2$O, ZnO and Bi$_2$O$_3$) glassy system. Similar observation was reported by G.H. Liao et al [25] in a gallo-germanate glass. Hence, by varying the glass composition, emission bandwidth can be tuned.

The concentration dependent broadening of the luminescence in Er^{3+} doped phosphate glass is also reported [26]. Fig. 5 shows the normalized emission spectra of the Er^{3+} doped glass samples measured under excitation at 976 nm. With the increase of erbium concentration from 1 mol% to 2 mol%, the FWHM of this band slightly increases from 30 to 39 nm. Similar observation was reported for tellurite glasses [22, 27 and 28]. According to M. Reza Dousti [22], such broadenings are associated to the radiative trapping which is more likely in a 3-level system comprising overlapped absorption and emission spectra (self-absorption process). In fact, at higher concentrations of erbium ions, the short ion-ion distance facilitates the self-absorption process yielding the broadening in detected NIR emission.

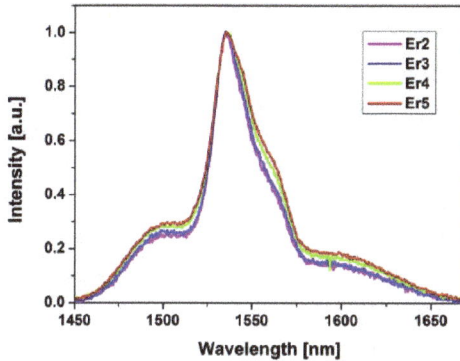

Figure 5. NIR emission spectra of phosphate glass samples for different Er^{3+} concentrations obtained under excitation at 976 nm.

Besides these physical origins for the broadening of the 1.53 μm emission, co-doping with second rare earth elements such as ytterbium ions (Yb^{3+}) and praseodymium (Pr^{3+}) or metallic nanoparticles (NPs) (generally used to improve the pump absorption) also leads to larger inhomogeneous broadening of the emission spectra.

The following paragraphs of this chapter contain a general description of the effect of co-doped with RE and silver nanoparticles (Ag NPs) in the 1.53 μm emission of the Er^{3+} doped phosphate glasses for use in broadband applications.

3. Lanthanides dependent broadening

Intense studies have been carried out upon Er^{3+} doped phosphate glasses in order to optimize the gain and the emission cross-section of the $^4I_{13/2} \rightarrow {}^4I_{15/2}$ transition of Er^{3+} and to examine their suitability as potential optical glasses for broadband applications such as Erbium-doped fiber amplifiers (EDFAs). In fact, the most widespread amplifier is erbium-doped fiber amplifier (EDFA) operating at 1530 nm wavelength, which has been utilized for the C-band (1530–1565 nm) optical communications [29, 30]. However, the conventional EDFAs are difficult to meet the optical communication requirements with large transmission capacity at fast rates as their gain bandwidth has no more than 70 nm

(1530–1600 nm) [31]. Hence, an advisable extension of EDFAs would be the addition of other RE such as ytterbium ions (Yb^{3+}) and praseodymium (Pr^{3+}).

We provide the following short synopsis germane to study the effect of the addition of other RE in the 1.55 μm emission of the Er^{3+} doped phosphate glasses

3.1   Ytterbium (Yb^{3+}) ions sensitizing effect

In general, the two transitions of the Er^{3+}: $^4I_{15/2} \rightarrow {}^4I_{11/2}$ (at 980 nm) and $^4I_{15/2} \rightarrow {}^4I_{9/2}$ (at 800 nm) can be improved by adding Yb^{3+} as a sensitizer since the Yb^{3+} ions exhibit an intense and broad absorption cross-section between 850 and 1080 nm. Besides, Yb^{3+} ion has high quantum yield, strong mission intensity in the NIR domain and long fluorescence lifetime, which provides an excellent complement to achieve the broadband NIR emission band of Er^{3+} ions.

Recently, the emission properties of Er^{3+}/Yb^{3+} co-doped phosphate glasses have been discussed in different phosphate glass compositions. For example, the effect of Yb^{3+} ions on the luminescence of Er^{3+} doped zinc phosphate glasses is reported by A. Langar et al. [13]. It was found that addition of Yb^{3+} enhances the luminescence intensity of the $^4I_{13/2} \rightarrow {}^4I_{15/2}$ transition. They concluded that by increasing the amount of Yb^{3+} in (74-y) NaH_2PO_4–$20ZnO$–$5Li_2CO_3$–$0.5Er_2O_3$–yYb_2O_3 (where y=0, 0.5, 1 mol %) glassy system. Moreover, the values of decay times support the increase in luminescence intensity of $^4I_{13/2} \rightarrow {}^4I_{15/2}$ transition. In fact, the $^4I_{13/2}$ lifetime increases with increasing Yb^{3+} load from 0 to 1 mol%. The observed fluorescence enhancement as well as the increase in the lifetime of the emitting level $^4I_{13/2}$ was explained by energy transfer from Yb^{3+} to Er^{3+} ions under the excitation of 980 nm. To illustrate the effect of Yb^{3+} in the emission of the Er^{3+} doped phosphate glasses, Fig.6 compares some normalized emission spectra for several Yb^{3+} concentrations. More extensive data are provided in Table 1 which lists measured full width at half maximum for compositions representing most of phosphate glasses types suitable for optical applications.

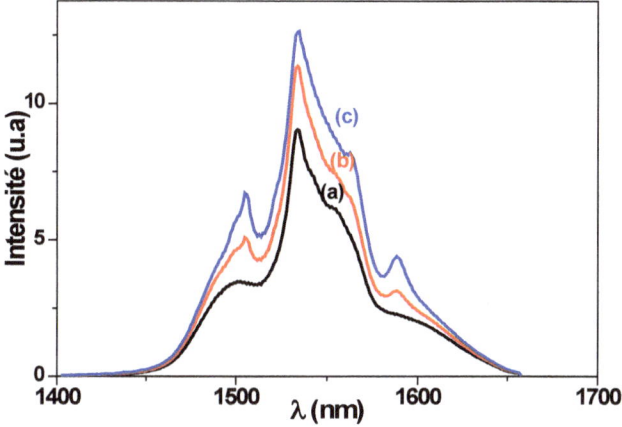

Figure 6: PL spectra relatives to $^4I_{13/2} \rightarrow ^4I_{15/2}$ transition, under 976 nm excitation wavelength, of NZLE0 (a), NZLE1 (b) and NZLE2 (c) glasses. Figure is adopted from Ref. [13]

As can be seen in Table 1, $\Delta\lambda_{eff}$ increases from 65 to 70 nm and the FWHM increases from 43 to 65 nm with increasing Yb^{3+} load from 0 to 1 mol%. These values are larger than those reported for materials commonly used for optical amplifiers such as phosphate (37 - 44 nm) [32, 33] and silicate glasses (40 nm) [34], while in the same order of magnitude for Er-Yb doped tellurite glasses (54–63 nm) [35]. According to authors [13], the broadening of the luminescence band in this glass is due to the variation of local structure and coordination numbers surrounding Er^{3+} ions site. In fact, the interaction between Er^{3+} and Yb^{3+} ions strongly influences the local structure and coordination numbers surrounding Er^{3+} ions site. The sites to sites variation brings to an inhomogeneous broadening, which is clearly desirable for a broadband EDFA glass host. These results suggesting that the increase of Yb_2O_3 concentration effectively affects the differences in the ligand field of Er^{3+} from site to site and thus leads to larger inhomogeneous broadening of the emission spectra. Similar observation was reported by S. Hraiech et al. [36] in a phosphate glass. In another study, N. Sdiri et al. [7] showed in a phosphate-borate glass that the increase in the concentration of the Yb^{3+} ions improves the quantum efficiency of the broadband luminescence.

Table 1 : The effective width $\Delta\lambda_{eff}$, FWHM, emission cross-section (σ_e) and FWHM×σ_e of $^4I_{13/2} \rightarrow {}^4I_{15/2}$ transition. (All data taken from [13]).

Er^{3+}/Yb^{3+} (mol. %)	0.5/0	0.5/0.5	0.5/1	Aluminosilicate glass [37]	PBGG [38]
FWHM (nm)	43.842	44.839	65.764	43	38
$\Delta\lambda_{eff}$ (nm)	65.411	65.177	70.025	-	-
σ_e (x10^{-21} cm^2)	7.7	6.2	7.2	5.7	8.9
σ_e x FWHM (x10^{-21} nm.cm^2)	337.5	278	473.5	245.1	338.5

In addition, emission cross-section σ_e and the FWHM are important parameters in an optical amplifier's achieving broadband and high-gain amplification. The properties of the optical amplifier can be evaluated from the bandwidth quality factor FWHM×σ_e. Larger values for the FWHM×σ_e product and longer experimental lifetime's τ_{mes} imply wider gain bandwidth and lower pump threshold power. Table 1 shows the FWHM×σ_e of Er^{3+} in various glasses for comparison of the emission properties of the $^4I_{13/2} \rightarrow {}^4I_{15/2}$ transition. It can be seen that phosphate glasses co-doped with low Er^{3+}/Yb^{3+} has better bandwidth properties than other glass hosts such as Alumino-silicate glass (245.1 x10^{-21}nm.cm^2) [37] and PBGG (338.5 x10^{-21} nm.cm^2) [38], reported in published papers for EDFA. Furthermore, the bandwidth quality factor increased with the increase of Yb$_2$O$_3$ concentration. The large lifetimes of $^4I_{13/2}$ level as well as the higher bandwidth quality factor make 1.53µm fluorescence much stronger in the Er^{3+}/Yb^{3+} co-doped phosphate glasses than that in Er^{3+} singly doped glasses. This fact affirms that Yb^{3+} ions are highly advantageous to achieving a broadband EDFA glass host.

The effect of Yb^{3+} ions in the 1.53 µm emission of erbium-doped phosphate glasses can be discussed in terms of gain cross-section. Assuming that the population of Er^{3+} ions is distributed only between the $^4I_{15/2}$ ground state and the $^4I_{13/2}$ first excited state, the optical gain properties are directly associated with the absorption and emission cross-sections. From the measured absorption cross-section and the calculated emission cross-section by the Mc Cumber expression (see eq. 3), the gain cross-section can be evaluated as a function of the population inversion P using the following relation [39]:

$$G((\lambda) = N_{Er}[P\sigma_e(\lambda) - (1-P)\sigma_a(\lambda)] \tag{3}$$

where N_{Er} stands for the 1.0 mol% of Er^{3+} ions concentration and P represents the population inversion, defined as $P = (N_1/N_1 + N_2)$ in which N_1 and N_2 are the number of ions in the ground ($^4I_{15/2}$) and excited ($^4I_{13/2}$) states, respectively.

F. Rivera-López et al. [40] reported an experimental research on gain properties in Er^{3+}/Yb^{3+} co-doped phosphate glasses. They concluded that the energy-transfer process from Yb^{3+} ions to Er^{3+} ions makes co-doped phosphate glass an excellent active material for improving the amplification gain of the $^4I_{13/2} \rightarrow ^4I_{15/2}$ transition at around 1536 nm. Fig. 7 shows the wavelength dependence of the gain coefficient for the $^4I_{13/2} \rightarrow ^4I_{15/2}$ emission transition calculated for 1 mol% of Er_2O_3 and 1 mol% of Yb_2O_3.

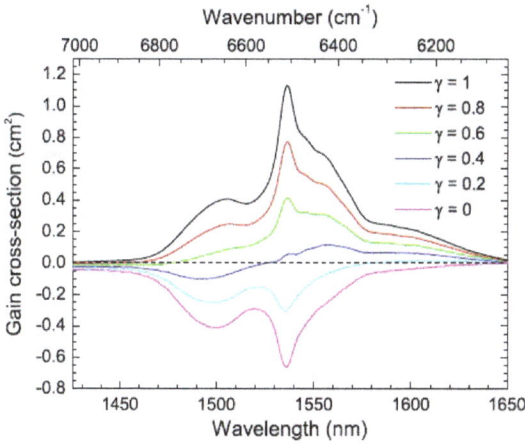

Figure 7: Gain cross-section in the eye-safe range for different values of the inversion of the population for the phosphate glass doped with 1 mol% of Er_2O_3 and 1 mol% of Yb_2O_3. Figure is adopted from Ref. [40]

It can be seen that the gain will be positive when the population inversion is larger than 0.4. Thus, it can be concluded that for a normal population inversion above 40%, the sample has a flat gain bandwidth in the range from 1475 to 1625 nm. They found that the gain cross-section is broader by around 30 nm more than the value found in conventional silicate-based EDFA and covers the C (1530–1565 nm) and the L (1565–1625 nm) bands in the optical communication window, which means that more channels may be allowed in the WDM network. Hence, the presence Yb^{3+} ions as sensitizer its important role for 1.53 µm band broad and high-gain erbium-doped fiber amplifiers (EDFA).

3.2 Praseodymium (Pr^{3+}) ions effect

Praseodymium (Pr^{3+}) shows up promising to achieve some novel near-infrared emissions due to its rich multiple energy levels. In fact, the transitions $^1G_4 \rightarrow ^3H_5$ at 1330 nm [41] and $^3F_3, ^3F_4 \rightarrow ^3H_4$ at 1600 nm [42] had been widely investigated in a variety of hosts and therefore, optical amplification based on these transitions have been demonstrated. Praseodymium (Pr^{3+}) doped glasses may still exhibit another interesting emission in the NIR, due to the $^1D_2 \rightarrow ^1G_4$ transition. The broadband emission around 1500 nm that was attributed to this radiative decay was observed in ZBLAN [43], tellurite [44] and silicate [45] glasses. Nowadays, the interest in broadband optical amplifiers has continually motivated the research into NIR emitters. Recent studies have been carried out about the so called "superbroadband" emission of Pr^{3+} in single doped low phonon hosts such as bismuth–gallate [46], fluor-tellurite [47], tellurite–germanate [48] and boro-phosphate [49] glasses. The full-width at half maximum (FWHM) for the $^1D_2 \rightarrow ^1G_4$ transition reported is 140 nm in these glasses, simultaneously covering the E, S, C and L-bands as well. Hence, Pr^{3+} and Er^{3+} ions are the perfect combination behind the reason that the $^1D_2 \rightarrow ^1G_4$ transition of Pr^{3+} at around 1481 nm provides an excellent complement to Er^{3+}.

Although there are not many reports on the effect of Pr^{3+} on the luminescence of Er^{3+} doped glasses, some research groups investigated the effect of Pr^{3+} ions on such systems. Recently, G.S. Li et al. [50] reported for the first time the near infrared (NIR) emission spectra of Pr^{3+}/Er^{3+} co-doped phosphate glasses with 483 nm excitation. The results (Fig.8) indicate a broadband emission extending from 1450 to 1640 nm , which covers the whole O, E, S, C, L and U bands.

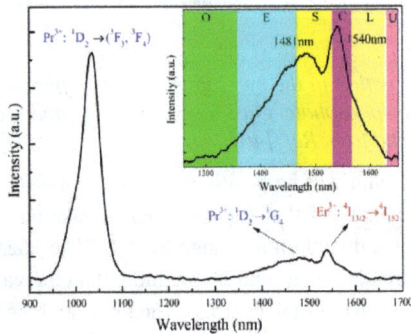

Figure 8: NIR emission spectrum of 0.3 mol% Pr^{3+} and 0.1 mol% Er^{3+} co-doped phosphate glasses under 483 nm excitation. The inset shows the details of emission spectra in wavelength region 1260–1650 nm. Figure is adopted from Ref. [50]

This superbroadband luminescence is contributed by the Er^{3+}: $^4I_{13/2} \rightarrow ^4I_{15/2}$ and Pr^{3+}: $^1D_2 \rightarrow ^1G_4$ transitions that leads to emissions located at 1540 and 1481 nm, respectively. This admirable result implies that Er^{3+}/Pr^{3+} co-doped phosphate glass fiber amplifiers operating at the O, E, S, C, L and U bands brings the possibility of effective signal amplification within broadband wavelength region. Based on this study, we conclude that addition of Pr^{3+} ions can be promising for broadband application especially for Erbium-doped fiber amplifiers (EDFAs). To the best of knowledge, there are many important parameters which are relevant for optical amplifiers for $^4I_{13/2} \rightarrow ^4I_{15/2}$ transition to avoid any mislead and mix-up such as effective bandwidth ($\Delta\lambda_{eff}$), full-width at half-maximum (FWHM), emission cross-section (σ_e), bandwidth quality factor (FWHM$\times\sigma_e$) and gain cross-section ($G(\lambda)$) which are important parameters for the Er^{3+}-doped fiber amplifier (EDFAs).

3.3   Metallic nanoparticles effect

In recent years, glasses containing both metallic nanoparticles (NPs) and rare-earth (RE) have been the subject of increasing interest due to their prospective utilization for photonic applications [51-55]. Special attention has been given to silver nanoparticles (Ag NP) due to their remarkable optical properties owing to the surface plasmon resonance (SPR) in the visible range such as silver gold or copper [52, 54]. In fact, the phenomenon of surface plasmon resonance is classically described as the oscillation of the free electrons with respect to the ionic background of the nanoparticle, when they are collectively excited by laser irradiation. The strong absorption cross-section related to the surface plasmon excitation in noble-metal nanoparticles and/or the large local field enhancement that is generated around the excited nanoparticle, make it possible to use such metal nanoparticles for the enhancement of the luminescence intensity emitted by rare-earth ions when present, in close vicinity of plasmonics nanoparticles [56, 57].

To the best of our knowledge, only few studies are found in literature with enhancement in 1.53 μm band fluorescence when embedding metallic NPs inside the glass host [8, 9, 58, 59, and 60].

Our recent extensive studies on Er^{3+} doped tellurite glasses reveal the potentiality of these metallic NPs for an efficient 1.53 μm broadband optical amplifiers [9]. We have reported the structural, spectroscopic and emission properties of Er^{3+} ions doped tellurite glasses in the presence of Ag NPs by varying $AgNO_3$ concentration. The simultaneous influence of the Ag NPs$\rightarrow Er^{3+}$ energy transfer and the contribution of the intensified local field effect due to the silver NPs, give origin to the enhancement of both the Photoluminescence (PL) intensity and the PL lifetime relative to the $^4I_{13/2} \rightarrow ^4I_{15/2}$ transition, whereas the quenching is ascribed to the energy transfer from Er^{3+} ions to silver NPs. Moreover, the results indicate a broadband emission extending from 1300 to 1640 nm. The broadening of the luminescence band in these glasses is mainly due to the interaction between Er^{3+} and Ag NPs strongly influencing the local structure and coordination numbers surrounding Er^{3+} ions site. In fact, we believe that $AgNO_3$ when present in the TeO_2 glass network produces a range of structural sites for RE ions. These structural sites could be a variant of trigonal bipyramid for TeO_4, trigonal pyramid for TeO_3 or a polygonal structure for $TeO_{3+\delta}$. The sites to sites variation brings to an inhomogeneous broadening which is clearly desirable for a broadband EDFA glass host. In another study, Rivera et al. [58] investigated the Er^{3+} doped tellurite glasses containing silver NPs and heat-treated above the glass transition. The small SPR peaks were revealed in an Er^{3+}/Ag NPs doped tellurite glass centered at 479 and 498 nm for 3 and 6 h annealed samples, respectively. Slight increase in FWHM and intensity of broadband emission of Er^{3+} ion (~1.55 μm) were observed by increasing the annealing time interval. Furthermore, the effect of annealing time on the emission properties of $^4I_{13/2} \rightarrow {}^4I_{15/2}$ transition has been investigated by H. Fares et al [8]. The presence of silver NPs with size of about 20-45 nm is confirmed by TEM. The PL intensity and PL lifetime relatives to the $^4I_{13/2} \rightarrow {}^4I_{15/2}$ transition are found to be enhanced. Local field enhancement induced by silver NPs is found to be responsible for enhancement, while the energy transfer from Er^{3+} ions to Ag NPs and/or the conglomerates of silver NPs are responsible factors for quenching of the emission intensity and PL lifetime of Er^{3+}: $^4I_{13/2}$ level. In addition, the results indicate that $\Delta\lambda_{eff}$ as well as FWHM increase after 10 h of heat-treatment. As mentioned before, the broadening of the luminescence band in these glasses is mainly due to the interaction between Er^{3+} and Ag NPs which strongly influences the local structure and coordination numbers surrounding Er^{3+} ions site. We believe that the increase in particle size of Ag NPs with the annealing time affects the differences in the ligand field of Er^{3+} from site to site and thus leads to larger inhomogeneous broadening of the emission spectra.

Since there are no studies reported on the effect of Ag NPs on the luminescence properties of Er^{3+} doped phosphate glasses for an efficient 1.53 μm broadband optical amplifiers, we have recently investigated the effect of Ag NPs on spectroscopic properties of Er^{3+} embedded in phosphate glass [12]. The TEM images confirmed the formation of spherical silver NPs having an average diameter in the range of 20-40 nm inside the glass matrices (Fig. 9 (a)). A broad absorption band was observed around 403 nm due to the surface Plasmon resonance (SPR) of Ag NPs (Fig. 9 (b)).

It was found that the presence of silver NPs improves the photoluminescence (PL) intensity and the PL lifetime relatives to the $^4I_{13/2} \rightarrow {}^4I_{15/2}$ transition. Maximum enhancement in the PL intensity of 1.53 μm band was obtained in sample doped 1 mol% Ag_2CO_3. Such enhancements are attributed to the strong local electric field induced by

SPR of silver NPs as the major factor. The energy transfer from the metal to rare earth ions is mainly proposed to be an inefficient factor for luminescence enhancement. Therefore, some energy transfer from 980 nm pump light absorbed by Ag NPs to Er^{3+} ions is possible. However, $\Delta\lambda_{eff}$ and FWHM decrease with the addition of silver NPs. These results are not fully understood and need rethinking in order to ascertain its origins.

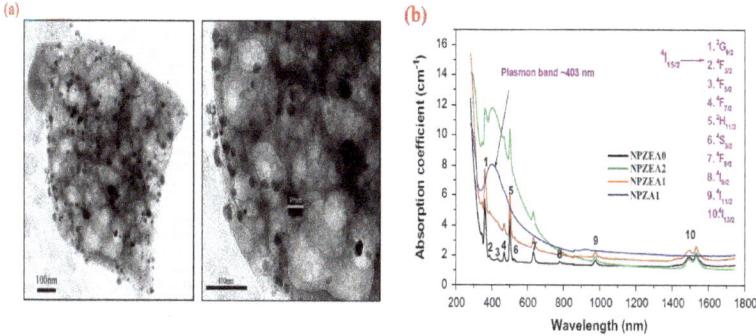

Figure 9: (a) TEM images of phosphate glass sample doped 1 mol% Ag_2CO_3. (b) Absorption spectra of Ag/Er^{3+} doped phosphate glasses. Figures are adopted from Ref. [12]

The effect of annealing temperature on the spectroscopic properties of Er^{3+} doped phosphate glasses is also investigated [61]. The SPR band shows a disordered shift in the range of 420-570 nm. The intensity of the SPR band increases with the annealing time caused by the nucleation of silver nanoparticles. The enhancement in the luminescence of Er^{3+} doped phosphate glass by introduction of silver NPs is also observed (Fig.10). we showed that under 488 nm excitation wavelength, maximum enhancement of the order of 3 times is observed after 12 h of heat-treatment compared with the un-annealed glass, while further annealing resulted in quenching of the emission intensity. This is obviously due to the increase in the concentration of NPs at higher annealing temperature and afterwards the average distance Ag-Ag NPs is smaller and hence increases the local field with annealing time until it is saturated. Therefore, one of the reasons for the enhancement could be the local field effect (LFE) which is induced by SPR of metallic NPs. Another proposed reason for the enhancement in luminescence intensity could be the energy transfer from silver NPs to Er^{3+} ions (Ag→Er^{3+}). On the other hand, quenching is attributed to the increased proximity between the Er^{3+} ions and silver NPs as a consequence of the growth in the amount of NPs (small metallic NPs aggregates or clusters).

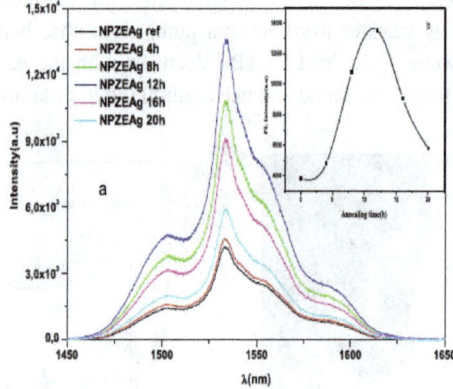

Figure 10: The effect of annealing time on the emission intensity.

More important, Absorption cross-section (σ_a), calculated emission cross-section (σ_e), the effective bandwidth (FWHM), emission bandwidth ($\Delta\lambda_{eff}$) and the bandwidth quality factor (FWHM$\times\sigma$e) for the $^4I_{13/2} \rightarrow {}^4I_{15/2}$ transition, were determined and compared to other glasses. As expected, the emission cross-section (σ_e) supports the increase in luminescence intensity of $^4I_{13/2} \rightarrow {}^4I_{15/2}$ transition in prepared glasses and confirms the mechanism of energy transfer from silver NPs to Er^{3+} ions. It has been observed also that the ($\Delta\lambda_{eff}$) increased from 52.87 to 69.58 nm and that the FWHM increased from 42.12 to 58.17 nm after 12 h of heat-treatment [61]. The broadening of the luminescence band in these glasses is mainly due to the interaction between erbium ions and silver NPs. On the basis of these results, bandwidth quality factor (FWHM$\times\sigma_e$) was calculated and the results indicate that FWHM$\times\sigma_e$ relative to NPZEAg12h after 12 h of heat-treatment is much larger than those reported in published papers for EDFA, which indicates that the NPZEAg12h sample glass is a favorable material for optical amplifiers.

4. Conclusion

The aim of this chapter was to revisit the current achievements on the state of art of understanding the origins of the broadening of the broadband emission of Er^{3+} doped phosphate glasses for broadband applications such as Erbium-doped fiber amplifiers (EDFAs). Until now, intense studies have been carried out upon Er^{3+} doped phosphate glasses in order to optimize the gain and the emission cross-section of the $^4I_{13/2} \rightarrow {}^4I_{15/2}$

transition of Er^{3+} and to examine their suitability as potential optical glasses for broadband applications. Incorporation of larger concentrations of RE ions, introduction of second dopant, thermal treatments, different synthesizing methods and varying the glass host matrix are among the common techniques in order to modify the environment of the RE ions, which significantly can alter its optical properties. In this chapter, we focused on the effect of the addition of other RE as well as silver nanoparticles (Ag NPs) in the 1.55 μm emission of the Er^{3+} doped phosphate glasses for use in broadband applications.

References

[1]   M. Yamane and Y. Asahara. Glasses for Photonics. Cambridge - University Press, Cambridge, United Kingdom; 2002.

[2]   V.A.G. Rivera, E.F. Chillce, E.G. Rodrigues, C.L. Cesar, L.C. Barbosa. Proc. SPIE 2006; 6116: 190-193. http://dx.doi.org/10.1117/12.647152

[3]   V.A.G. Rivera, E.F. Chillce, E.G. Rodrigues, C.L. Cesar, L.C. Barbosa. J. Non-Crys. Sol. 2006; 353: 125-130

[4]   P.J. Mears, L. Reekie, I.M. Jauncey and D.N. Payne. Elec. Lett, 23 (1987) 1026.http://dx.doi.org/10.1049/el:19870719

[5]   E. Desurvire, R.J. Simpson and P.C. Becker. Opt. Lett,12 (1987)888. http://dx.doi.org/10.1364/OL.12.000888

[6]   E. Desurvire. Phys. Today, 47 (1994) 20. http://dx.doi.org/10.1063/1.881418

[7]   N. Sdiri, H. Elhouichet, C. Barthou, and M. Ferid, J. Mol. Struct. 1010, (2012) 85. http://dx.doi.org/10.1016/j.molstruc.2011.11.036

[8]   H. Fares, H. Elhouichet, B. Gelloz, and M. Férid, J. Appl. Phys. 116, (2014) 123504. http://dx.doi.org/10.1063/1.4896363

[9]   H. Fares, H. Elhouichet, B. Gelloz, and M. Férid, J. Appl. Phys. 117, (2015) 193102. http://dx.doi.org/10.1063/1.4921436

[10] W. Stambouli, H. Elhouichet, C. Barthou, M. Férid, J. Alloys Compd, 580 (2013) 310. http://dx.doi.org/10.1016/j.jallcom.2013.06.115

[11] J. Qi, T. Xu, Y. Wu, X. Shen, S. Dai, and Y. Xu, Opt. Mater, 35 (2013) 2502. http://dx.doi.org/10.1016/j.optmat.2013.07.009

[12] I. Soltani, S. Hraiech, K. Horchani-Naifer, H. Elhouichet, M. Férid, Opt. Mater, 46 (2015) 454. http://dx.doi.org/10.1016/j.optmat.2015.05.003

[13]   A. Langar, C. Bouzidi, H. Elhouichet, and M. Férid, J. Lumin, 148 (2014) 249.
        http://dx.doi.org/10.1016/j.jlumin.2013.12.008

[14]   Chen B Y et al Chin. Phys. Lett, 20 (2003) 2056. http://dx.doi.org/10.1088/0256-
        307X/20/11/044

[15]   Amjad R J et al Chin. Phys. Lett, 29 (2012) 087304.
        http://dx.doi.org/10.1088/0256-307X/29/8/087304

[16]   Raja J. Amjad, M. R. Sahar, S. K. Ghoshal, M. R. Dousti, S. Riaz, A. R. Samavati,
        M. N. A Jamaludin, S. Naseem, CHIN. PHYS. LETT. Vol. 30, No. 2 (2013)
        027301

[17]   M. J. Weber, J. D. Myers, D. H. Blackburn, J. Appl. Phys, 52 (1981) 2944.
        http://dx.doi.org/10.1063/1.329034

[18]   D. E. Mc Cumber, Phys. Rev. 136 (4A), A954 (1964).
        http://dx.doi.org/10.1103/PhysRev.136.A954

[19]   Y.D. Huang, M. Mortier, F. Auzel, Opt. Mater, 15 (2001) 243.
        http://dx.doi.org/10.1016/S0925-3467(00)00039-2

[20]   A. Kenyon, Recent Developments in Rare-earth Doped Materials for
        Optoelectronics, 2002.

[21]   J. Coelho, J. Azevedo, G. Hungerford, N.S. Hussain, Opt. Mater, 33 (2011) 1167.
        http://dx.doi.org/10.1016/j.optmat.2011.02.003

[22]   M. Reza Dousti, Journal of Molecular Structure, 1100 (2015) 415.
        http://dx.doi.org/10.1016/j.molstruc.2015.07.063

[23]   Joanna Pisarska, Wojciech A. Pisarski, Tomasz Goryczka, Radosław Lisiecki,
        Witold Ryba-Romanowski, Journal of Luminescence, 160 (2015) 57.
        http://dx.doi.org/10.1016/j.jlumin.2014.11.029

[24]   S. Babu, M. Seshadri, V. Reddy Prasad, Y.C. Ratnakaram, Materials Research
        Bulletin, 70 (2015) 935. http://dx.doi.org/10.1016/j.materresbull.2015.06.033

[25]   D.M. Shi, Y.G. Zhao, X.F. Wang, G.H. Liao, C. Zhao, M.Y. Peng, Q.Y. Zhang,
        Physica B, 406 (2011) 632.[26] Diego Pugliese, Nadia G. Boetti, Joris Lousteau,
        Edoardo Ceci-Ginistrelli,Elisa Bertone, Francesco Geobaldo, Daniel Milanese, J.
        Alloys Compd, 657 (2016) 678. http://dx.doi.org/10.1016/j.jallcom.2015.10.126

[27]   N. Jaba, H.B. Mansour, B. Champagnon, Opt. Mater, 31 (2009) 1242.
        http://dx.doi.org/10.1016/j.optmat.2009.01.006

[28]   S. Dai, C. Yu, G. Zhou, J. Zhang, G. Wang, L. Hu, J. Lumin, 117 (2006) 39.
       http://dx.doi.org/10.1016/j.jlumin.2005.04.003

[29]   K. Linganna, K. Suresh, S. Ju, W.T. Han, C.K. Jayasankar, V. Venkatramu, Opt.
       Mater. Express, 5 (2015) 1689. http://dx.doi.org/10.1364/OME.5.001689

[30]   Z.Y. Zhao, B. Ai, C. Liu, Q.Y. Yin, M.L. Xia, X.J. Zhao, Y. Jiang, J. Am. Ceram.
       Soc, 98 (2015) 2117. http://dx.doi.org/10.1111/jace.13592

[31]   R.R. Xu, Y. Tian, M. Wang, L.L. Hu, J.J. Zhang, Opt. Mater. 33 (2011) 299–302.
       http://dx.doi.org/10.1016/j.optmat.2010.09.004

[32]   S. Jiang, T. Luo, B.-C. Hwang, F. Smekatala, K. Seneschal, J. Lucas, N.
       Peyghambarian, J. Non-Cryst. Solids 263-264 (2000) 364.
       http://dx.doi.org/10.1016/S0022-3093(99)00646-8

[33]   F. Rivera-López, et al., Opt. Mater, 34 (2012) 1235.
       http://dx.doi.org/10.1016/j.optmat.2012.01.017

[34]   X. Zou, T. Izumitani, J. Non-Cryst. Solids, 162 (1993) 68.
       http://dx.doi.org/10.1016/0022-3093(93)90742-G

[35]   I. Jlassi, H. Elhouichet, M. Ferid, C. Barthou., J. Lumin, 130 (2010) 2394.
       http://dx.doi.org/10.1016/j.jlumin.2010.07.026

[36]   S. Hraiech, M. Férid, Y. Guyot, G. Boulon, J. Rare Earths, 31 (2013) 685.
       http://dx.doi.org/10.1016/S1002-0721(12)60343-3

[37]   A. Polman, J. Appl. Phys, 82 (1997) 1. http://dx.doi.org/10.1063/1.366265

[38]   G.F. Yang, D.M. Shi, Q.Y. Zhang, Z.H. Jiang, J. Fluoresc, 18 (2008) 131.
       http://dx.doi.org/10.1007/s10895-007-0251-8

[39]   El Sayed Yousef, J. Alloys Compd, J. Alloys Compd, 561 (2013) 234.

[40]   F. Rivera-López, P. Babu, L. Jyothi, U. R. Rodríguez-Mendoza, I. R. Martín, C. K.
       Jayasankar, V. Lavín, Optical Materials, 34 (2012) 1235.
       http://dx.doi.org/10.1016/j.optmat.2012.01.017

[41]   Y. Ohishi, T. Kanamori, T. Kitagawa, S. Takahashi, E. Snitzer, G.H. Sigel, Opt.
       Lett, 16 (1991) 1747. http://dx.doi.org/10.1364/OL.16.001747

[42]   Y.G. Choi, K.H. Kim, B.J. Park, J. Heo, Appl. Phys. Lett, 78 (2001) 1249.
       http://dx.doi.org/10.1063/1.1350958

[43]   Y. Ohishi, T. Kanamori, T. Kitagawa, S. Takahashi, E. Snitzer, G.H. Sigel, Opt.
       Lett, 16 (1991) 1747. http://dx.doi.org/10.1364/OL.16.001747

[44]   S.Q. Man, E.Y.B. Pun, P.S. Chung, Opt. Commun, 168 (1999) 369.
        http://dx.doi.org/10.1016/S0030-4018(99)00374-0

[45]   Y.G. Choi, J.H. Baik, J. Heo, Chem. Phys. Lett, 406 (2005) 436.
        http://dx.doi.org/10.1016/j.cplett.2005.03.028

[46]   B. Zhou, E.Y. Pun, Opt. Lett, 36 (2011) 2958.
        http://dx.doi.org/10.1364/OL.36.002958

[47]   B. Zhou, L. Tao, Y.H. Tsang, W. Jin, E.Y.-B. Pun, Opt. Exp, 20 (2012) 3803.
        http://dx.doi.org/10.1364/OE.20.003803

[48]   X. Liu, B.J. Chen, E.Y.B. Pun, H. Lin, J. Appl. Phys, 111 (2012) 116101.
        http://dx.doi.org/10.1063/1.4722997

[49]   Q. Sheng, X. Wang, D. Chen, J. Lumin, 135 (2013) 38.
        http://dx.doi.org/10.1016/j.jlumin.2012.10.040

[50]   G.S. Li, C.M. Zhang, P.F. Zhu, C. Jiang, P. Song, K. Zhu, Ceramics International,
        42 (2016) 5558. http://dx.doi.org/10.1016/j.ceramint.2015.12.026

[51]   Sérgio P. A. Osorio, Victor A. Garcia Rivera, Luiz Antonio O. Nunes, Euclydes
        Marega, Danilo Manzani, Younes Messaddeq, Plasmonics, 7 (2012) 53.

[52]   M. Reza Dousti, M. R. Sahar, Raja J. Amjad, S. K. Ghoshal, A. Khorramnazari, A.
        Dordizadeh Basirabad, A. Samavati, European Physical Journal D, 66 (2012) 237.
        http://dx.doi.org/10.1140/epjd/e2012-30089-1

[53]   T. Som, B. Karmakar, J. Appl. Phys, 105 (2009) 013102.
        http://dx.doi.org/10.1063/1.3054918

[54]   T. Som, B. Karmakar, Applied Surface Science, 255 (2009) 9447.
        http://dx.doi.org/10.1016/j.apsusc.2009.07.053

[55]   Nehal Aboulfotoh, Yahia Elbashar, Mohamed Ibrahem, Mohamed Elokr,
        Ceramics International, 40 (2014) 10395.
        http://dx.doi.org/10.1016/j.ceramint.2014.02.125

[56]   A. Chiasera, M. Ferrari, M. Mattarelli, M. Montagna, S. Pelli, H. Portales, J.
        Zheng, G.C. Righini, Optical Materials, 27 (2005) 1743.
        http://dx.doi.org/10.1016/j.optmat.2004.11.044

[57]   Zahra Ashur Said Mahraz, M.R.Sahar, S.K.Ghoshal, M. R. Dousti, R. J. Amjad,
        Materials Letters, 112 (2013) 136.

[58] V.A.G. Rivera, S.P.A. Osorio, Y. Ledemi, D. Manzani, Y. Messaddeq, L.A.O. Nunes, E. Marega Jr, Opt. Express, 18 (2010) 25321. http://dx.doi.org/10.1364/OE.18.025321

[59] V.A.G. Rivera, Y. Ledemi, S.P.A. Osorio, D. Manzani, F.A. Ferri, Sidney J.L. Ribeiro, L.A.O. Nunes, E. Marega Jr, J. Non-Cryst. Solids 378 (2013) 126. http://dx.doi.org/10.1016/j.jnoncrysol.2013.07.004

[60] Yawei Qi, Yaxun Zhou, Libo Wu, Fengjing Yang, Shengxi Peng, Shichao Zheng, Dandan Yin, Journal of Luminescence, 153 (2014) 401. http://dx.doi.org/10.1016/j.jlumin.2014.03.069

[61] I. Soltani, S. Hraiech, K. Horchani-Naifer, H. Elhouichet, B. Gelloz, M. Ferid, J. Alloys Compd, 686 (2016) 556. http://dx.doi.org/10.1016/j.jallcom.2016.06.027

CHAPTER 3

Lanthanum doped borophosphate glasses for nuclear waste immobilization

Fu Wang, Qilong Liao

School of Material Science and Engineering, Southwest University of Science and Technology, Mianyang 621010, PR China

Abstract

Recently, lanthanum-contained iron borophosphate glasses have received comprehensive attention for high-level radioactive wastes immobilization purpose because this system glass possesses both high elementary loading for chlorides, sulfates and heavy metals and good thermal properties. Moreover, the irradiation stability and chemical durability are comparable to those of iron phosphate base glasses and widely used borosilicate glass waste forms. This chapter summarizes recent progress on the suitability of lanthanum-contained iron borophosphate for nuclear waste immobilization from the aspect of structure, chemical durability, irradiation stability, thermal properties, etc.

Keywords

Iron Borophosphate Glasses; Lanthanum; Nuclear Waste Immobilization; Irradiation Stability; Thermal Properties; Structural Stability

Contents

1.   Introduction

Immobilization of high-level nuclear wastes (HLWs) in a suitable matrix is still highly challenging. Vitrification is considered as one of the best methods for HLWs immobilization [1-8]. Borosilicate glasses are widely used vitreous matrixes. However, some HLWs waiting for disposal are rich in phosphates, chromium oxide and heavy metals. These constituents are poorly soluble in borosilicate glasses [2]. Therefore, it is necessary to look for an alternate vitrification matrix. Iron phosphate glasses are potential host materials because of their low melting temperature, high chemical durability and large compositional flexibility [4-8]. This glass system is a good host material for HLWs containing abovementioned constituents [6-9]. Previous investigations have been focused on $40Fe_2O_3–60P_2O_5$ (mol%) base glass. The glass shows the best chemical durability among several iron phosphate glasses [6, 8]. Waste components or simulated nuclides have been added to this composition to simulate the actual wastes. However, the irradiation stability and thermal properties of the iron phosphate system glasses are lower [10, 11].

Recently, there have been some investigations on boron containing iron phosphate glasses [10, 12-19]. In general, these investigations are focused on the effects of boron addition on the structure and thermal properties of the glasses [10, 12-14]. Thermal stability is one of the criteria for glassy waste forms. Because glassy waste forms need to resist crystallization during cooling, especially, when melting in industrial scale, this problem becomes more important. It is found that addition of boron improves the thermal stability of iron phosphate glasses [10, 12, 14]. Boron also has a large neutron absorption cross section. Thus, it is advantageous to have boron in glassy waste forms, which both increase the thermal and irradiation stability of the resultant waste forms. Moreover, the boron doping has insignificant effects on its chemical durability. The dissolution rates of boron containing bulk iron phosphate glasses are found to be comparable to those of base iron phosphate glasses [13]. These effects of B_2O_3 on the properties of iron phosphate glasses are favorable for HLWs immobilization.

It has been revealed that the solubility of HfO_2 in the structure of iron borophosphate glasses is about 2 mol% [15]. The performances of $36Fe_2O_3–10B_2O_3–54P_2O_5$ glass (IBP glass) immobilizing oxides or elements coming from the constituents of HLWs have been investigated by our group [16-18]. The results show that good chemical durability and structural stability still remain for this glass containing 20 mol% Na_2O/K_2O [16], 9 mol% CeO_2 which are considered as surrogates for nuclide Pu^{4+} [17] or appropriate content of ZrO_2 [18]. Especially, the solubility limit of the Ce element is shown to be 9 mol% in the $36Fe_2O_3-10B_2O_3-54P_2O_5$ glass [17]. Moreover, the aqueous dissolution rates of iron borophosphate glasses containing 30 wt.% simulated nuclear waste are comparable to widely used borosilicate waste forms [19].

Heavy metals are currently incorporated in borosilicate nuclear glass matrices. However, waste loading capacity is quite low and there is an increasing need for alternative glass hosts that can incorporate higher amounts of wastes. This is especially true for those wastes containing significant amounts of PuO_2, ZrO_2 or La_2O_3. In this chapter, the effect of La_2O_3 on the glass forming range, density, structure features and chemical durability of the IBP glass are discussed. This glasses system, lanthanum doped borophosphate glass, are introduced from the aspects of glass forming range, density, structure features and chemical durability as a potential host for disposal of high-level nuclear wastes.

Two series of compositions have been discussed [20]: in series A, the IBP glass has been directly doped with La_2O_3; in series B, only Fe_2O_3 in the starting composition has been replaced by La_2O_3. The samples are labeled by the La_2O_3-doping modes and the molar amount of La_2O_3. For example, in series A, La6 sample represents $36Fe_2O_3-10B_2O_3-54P_2O_5$ glass doped with 6 mol% La_2O_3. In series B, LF6 sample represents substituting of 6 mol% La_2O_3 for Fe_2O_3 in the composition of $36Fe_2O_3-10B_2O_3-54P_2O_5$ glass. The chemical compositions of all samples were examined by X-ray fluorescence spectroscopy and are listed in Table 1.

Table 1 *Chemical compositions of analysed glasses from La_2O_3 doped iron borophosphate compounds.*

Samples	Molar compositions (mol%)			
	Fe_2O_3	B_2O_3	P_2O_5	La_2O_3
IBP glass	35.62	9.53	54.85	0
La3	34.91	9.40	52.67	3.02
La6	33.83	9.21	50.91	6.05
La9	32.77	8.89	49.36	8.98
LF3	33.21	9.5	54.28	3.01
LF6	30.43	9.52	54.08	5.97
LF9	27.47	9.48	54.06	8.99

2. Properties investigation

2.1 Glass forming ability

XRD patterns of the samples in both series are shown in Fig. 1. The JCPDS standard card of $LaPO_4$ crystal (PDF# 32-0493) is inserted in the figures for analysis assistance. From Fig. 1, it is shown that the iron borophosphate glass samples containing 3 mol% La_2O_3, both in series A and B, are fully amorphous. For the iron borophosphate glass samples containing more than 6 mol% (including 6 mol%) La_2O_3, $LaPO_4$ crystalline phase (PDF *No.* 32-0493, space group: Monoclinic, $P21/n(14)$) is detected in their structure. Moreover, further increase in the content of added La_2O_3, the number and intensity of diffraction peaks assigned to $LaPO_4$ crystal increase. This indicates that La_2O_3 has stronger ability to make the glass crystallize. Because the radius of La^{3+} (1.06 Å) is larger than that of Fe^{2+} (0.61 Å) /Fe^{3+} (0.49 Å), the cation field strength, Z/r^2, of La^{3+} is smaller than that of Fe^{2+}/Fe^{3+}, leading to weaker glass forming ability of La_2O_3 [20]. Comparing Fig. 1(a) and (b), the intensity of peaks assigned to $LaPO_4$ crystal for the samples in series B is much higher than that of the samples in series A. It is also ascribed to the weaker glass forming ability of La_2O_3 and that the doped La_2O_3 resultantly facilitates the crystallization of the iron borophosphate glasses.

SEM of typical samples in series A and the EDS pattern of the crystalline phase are shown in Fig. 2. The figure demonstrates that the sample with 3 mol% La_2O_3 shows a pure glassy fracture surface. The sample with 6 mol% La_2O_3 begins to exhibit phase separation (crystal appears). And the amount of crystal phase increases with the increase of La_2O_3 content, but they still show a compact structure. The EDS pattern shows that the constituent elements of the crystal formed are La, P and O, and the molar ratios of La:P:O are about 17.76:16.87:60.10 (Fig. 2(d)) which are closed to the value of 1:1:4. By combining the XRD analysis results, the crystalline phase is determined to be $LaPO_4$. Fig. 3 shows element distribution maps of La9 sample. This figure demonstrates the distributions of the main elements, including La, in the structural phase of the sample. It is observed that oxygen, phosphorus, iron and boron are homogeneously distributed over the entire glass phase in the sample. In contrast, lanthanum is concentrated in the separated crystalline phase. Therefore, the lanthanum element is rich in the crystalline phase formed when the content of doped La_2O_3 exceeds its solubility limit in the base IBP glass.

71

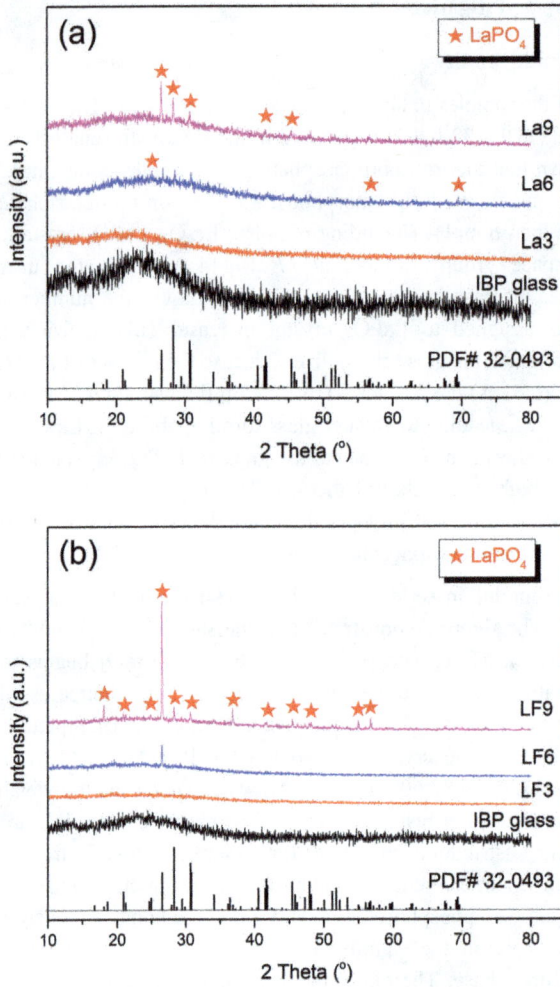

Fig. 1 *XRD patterns of the samples (a) in series A and (b) in series B.*

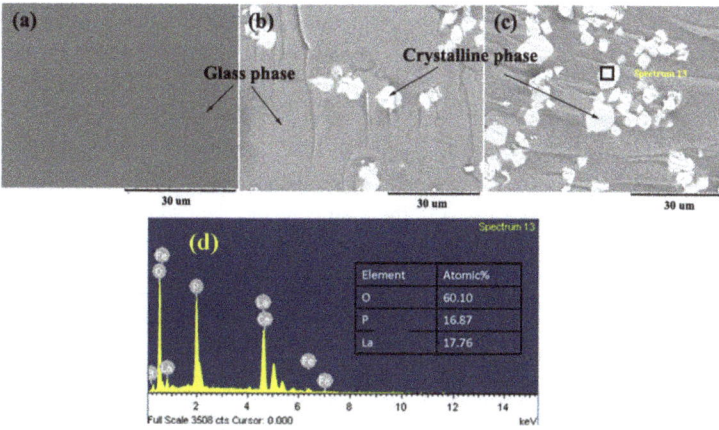

Fig. 2 *SEM of the sample (a) La3, (b) La6, (c) La9 and (d) EDS of the formed crystalline phase in samples.*

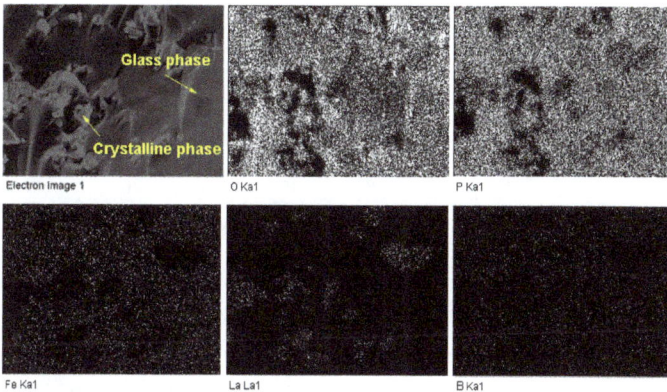

Fig. 3 *Element distribution maps of La9 sample.*

Combining the SEM, EDS, element distribution maps and XRD analysis, it can be concluded that the IBP glasses with 3 mol% La_2O_3 are fully amorphous. When 6 mol% or more La_2O_3 is added to the IBP glass, a $LaPO_4$ crystalline phase forms.

2.2 Densities

Fig. 4 shows the densities of the samples in both series. From the figure, it is shown that the densities of the samples change between 2.98 $g \cdot cm^{-3}$ and 3.42 $g \cdot cm^{-3}$. Moreover, the densities of the samples in both series increase with doped La_2O_3 content, which is mainly ascribed to the much higher atomic weight of La element compared to B, Fe and P elements. Especially, when a $LaPO_4$ crystalline phase forms, the densities increase more sharply because the $LaPO_4$ crystal formed shows a more compact structure compared with the base IBP glass. It also shows that the densities of the samples in series A are higher than those of the samples in series B because, for the glass/glass-ceramic containing the same amount of La_2O_3, the total content of La and Fe for the samples in series B are lower compared to the samples in series A.

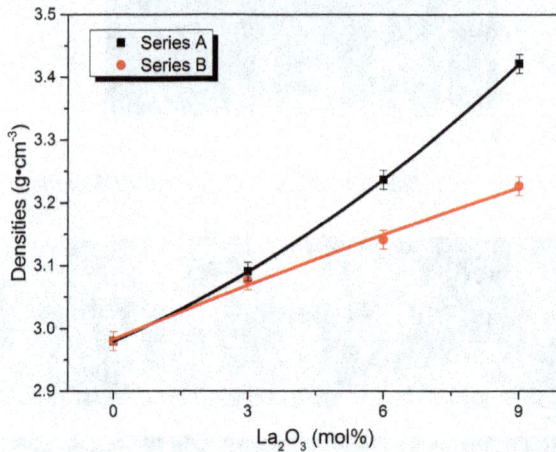

Fig. 4 *Densities of the La_2O_3-doped iron borophosphate glasses in both series.*

2.3   Glass transition temperature

DTA curves of the samples in series A are shown in Fig. 5. It is observed that all the curves are similar and characterized with an endothermic peak. The onset temperature of such an endothermic peak corresponds to the glass transition temperature (T_g). The T_g shifts to much higher values (from 523 °C to 578 °C) for the sample with 3 mol% La_2O_3 (Fig. 6) compared with the base IBP glass. According to the literature [21], the addition of lanthanum to aluminum phosphate glass strengthens the network by creating cross-links between phosphate chains, and the resultant increase in structural connectivity increases T_g. Similar T_g trend is observed for the iron borophosphate glass system. Moreover, although the base glass is depolymerized by the addition of lanthanum, the replacement of P–O–P bonds by La–O–P bonds contributes to the increase in T_g [22]. These results suggest that La acts a role of strengthening the cross-links between the phosphate units of the glasses [21, 22]. However, further increasing the content of doped-La_2O_3 (from 3 mol% to 9 mol%), T_g values decrease (from 587 °C to 530 °C). As revealed by XRD, the samples containing more than 6 mol% La_2O_3 contain a $LaPO_4$ crystalline phase. The formation of the $LaPO_4$ crystal will take away glass forming oxide P_2O_5. The decreased glass forming oxide weakens the glass structure, thus T_g decreases.

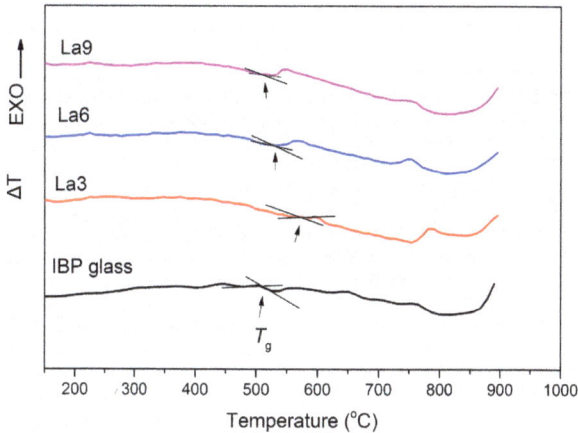

Fig. 5 *DTA curves of the samples in series A. The curves are arbitrarily offset for clarity.*

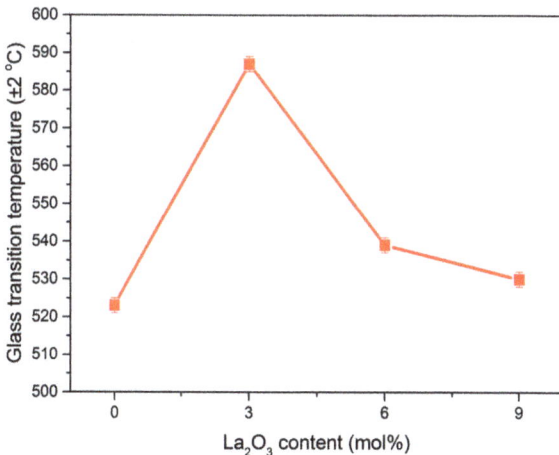

Fig. 6 *Glass transition temperature of the samples in series A as a function of La_2O_3 content.*

3.   Vibrational spectra

3.1   Infrared spectra

Fig. 7 shows the FTIR spectra of the IBP glasses containing La_2O_3 (between 2000 to 400 cm^{-1}). FTIR spectrum can provide information of molecular vibration or rotation associated with a specific bond or group, and is often used to investigate structural features of glassy materials. As shown in Fig. 7, the predominant absorption peaks are characterized by two peaks around 534 and 615 cm^{-1}, a quite weak peak around 751 cm^{-1}, broad overlapping bands in the 800–1350 cm^{-1} range, a peak around 1400 cm^{-1} with a shoulder at 1382 cm^{-1} and a peak around 1464 cm^{-1}. Some variations can be observed in the FTIR spectra of samples with different La_2O_3 content. The first noticeable variation occurs in the bands at 534 cm^{-1} and 615 cm^{-1}. The intensity of these bands increases with increasing La_2O_3 content. The second change is the appearance of a new shoulder band at ~953 cm^{-1} when the doping La_2O_3 is more than 6 mol%. In addition, the broad overlapping bands in the 800–1350 cm^{-1} range become stronger with increasing La_2O_3 content. These changes are similar for the samples in both series.

According to the literature (Table 2), the absorption band at ~ 534 cm^{-1} corresponds to the bending modes of O–P–O in $(P_2O_7)^{4-}$ groups [23, 24]. The band around 615 cm^{-1} is assigned to the stretching vibrations of Fe–O–P bonds [16, 25]. The band at 751 cm^{-1} is ascribed to the bending vibrations of B–O–B bonds in $[BO_4]$ groups [26]. It is known that broad bands are consistent with glass structures which accommodate many types of bonding state, and represent the overlap of vibrational absorbance for two or more bonding states [17, 18, 27]. In this paper, B_2O_3 is a glass former. When it is present in a composition together with P_2O_5, phosphate and borate groups will have broad overlapping bands in IR spectra in wavenumber between 800 and 1350 cm^{-1} [23, 24, 28, 29]. In Fig. 7, the intensity of the broad overlapping band becomes stronger with increasing La_2O_3. For samples containing a large amount of $LaPO_4$ crystal, the band becomes sharper. Usually, the more crystalline a material is, the more ordered its structure is. Therefore, the bands for a more crystalline material become sharper. In the figure, the sharper band for a partially crystallized material is found because the structural order of phosphate groups in $LaPO_4$ crystal within partially crystallized samples is higher compared with the fully amorphous base glass. Although vibrational modes of boron–oxygen triangular units ($[BO_3]$ and BO_2O^-) are located in range of 1360–1469 cm^{-1} [24], in this study, the weak absorption band at 1360–1469 cm^{-1} is also attributed to symmetric stretching vibrations of P=O bonds [23, 22].

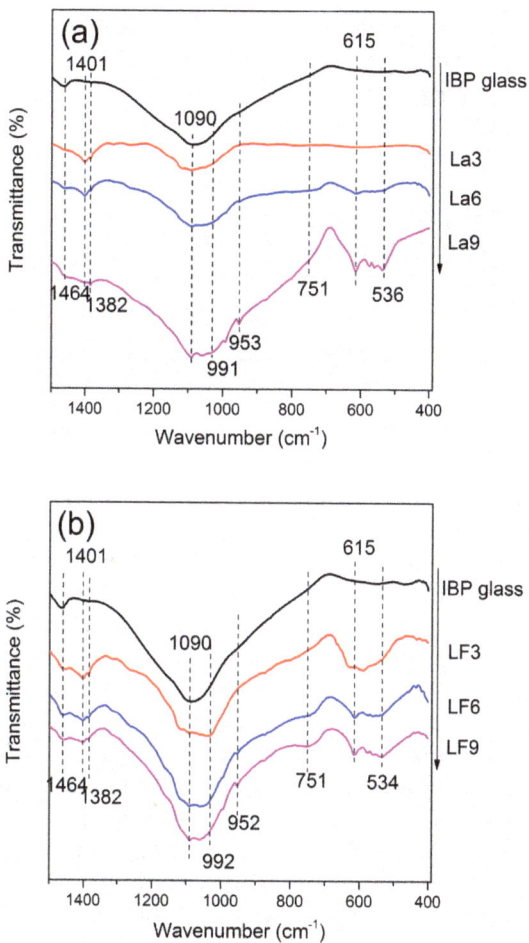

Fig. 7 *FTIR spectra of the La₂O₃ contained iron borophosphate glass samples, (a) in Series A and (b) in Series B.*

Table 2 *Assignments of FTIR spectra for the studied samples.*

Wavenumbers	Assignments	Ref.
~ 534 cm^{-1}	Bending modes of O–P–O in $(P_2O_7)^{4-}$ groups	[23, 24]
615 cm^{-1}	Stretching vibrations of Fe(La)–O–P bonds	[16, 25]
751 cm^{-1}	Bending vibrations of B–O–B bonds in [BO$_4$] groups	[26]
	Symmetric modes of P–O–P bonds in $(P_2O_7)^{4-}$ groups	
	P–O–B linkages	[23, 24]
800 - 1350 cm^{-1}	Stretching vibration of B–O bonds in [BO$_4$] units	[28, 29]
	Stretching vibrations of Q^0 groups	
	Asymmetric stretching of $(PO_3)^{2-}$ in Q^1 units	
~953 cm^{-1}	La–O–P bond in the formed LaPO$_4$ crystals	[22, 30]
1360-1469 cm^{-1}	Symmetric stretching vibrations of P=O bonds	[22, 23]
	Modes of [BO$_3$] and BO$_2$O$^-$	[24]

With the increasing content of La$_2$O$_3$ in the IBP glasses, the intensities of the bands at 534 cm^{-1} and 615 cm^{-1} increase in the FTIR spectra for the samples in both series. These show that the addition of La$_2$O$_3$ (glass modifying oxides) leads to the conversion of (PO$_3$)$^-$ metaphosphate groups to $(P_2O_7)^{4-}$ phosphate groups. In addition, La-O-P bond may form to replace part of P-O-P when La$_2$O$_3$ is added in these glasses. Due to the formed La–O–P bonds, the intensity of the band 615 cm^{-1} also increases and the glass structure is strengthened. The second feature observed in the IR spectra is the appearance of a new shoulder band at ~953 cm^{-1} when the doping La$_2$O$_3$ are more than 6 mol%. As La$_2$O$_3$ content increases, this band becomes sharp. Hence, this band should be related to the La bonding in the structure of glasses and LaPO$_4$ crystal [30]. In a recent study on hafnium containing iron borophosphate glasses, a band around this wavenumber is related to the formed HfP$_2$O$_7$ crystalline phase [15]. As revealed by XRD, the samples with more than 6 mol% La$_2$O$_3$ contain a LaPO$_4$ crystalline phase. Moreover, the wavenumber changes of the band at 925 cm^{-1} have been related to the La–O–P bond replacing P–O–P in iron phosphate glasses [22]. Therefore, the band at ~953 cm^{-1} arises from the formed LaPO$_4$ crystals in our studied IBP glasses. These results can also be deduced from the IR spectra of LaPO$_4$ [30].

3.2   Raman spectra

Raman spectra, in the wavenumber region between 1500 and 400 cm^{-1}, of the investigated samples in series A are shown in Fig. 8. In general, the spectra for the glasses have similar spectral features. For the base iron borophosphate glass, Raman peaks occur at ~ 465 cm^{-1}, ~ 575-629 cm^{-1}, ~ 963 cm^{-1}, ~ 1033-1008 cm^{-1} and ~ 1234 cm^{-1}. According to literature consensus, we have made the following band assignments for our spectra: ~ 465 cm^{-1} arises from O–P–O bending modes of Q^0 $(PO_4)^{3-}$ units [31, 32]. ~ 575-629 cm^{-1} is assigned to the vibrations of Q^1 $(P_2O_7)^{4-}$ units [33]. ~ 963 cm^{-1} is ascribed to asymmetric stretching of Q^0 $(PO_4)^{3-}$ monomer units [32, 34]. ~ 1033 cm^{-1} is attributed to symmetric stretching of Q^1 $(P_2O_7)^{4-}$ dimer units [35, 36]. ~ 1255 cm^{-1} is assigned to asymmetric stretching of Q^2 $(PO_3)^-$ metaphosphate units [32, 34]. From previous investigations of borophosphate glasses [37-39], it is known that borate groups show much lower Raman scattering efficiency than phosphate species and as the ratio of B/P is low in these glasses, therefore, there is no clear evidence for borate vibrations in the obtained Raman spectra. The main assignments of Raman spectra for the studied samples are listed in Table 3.

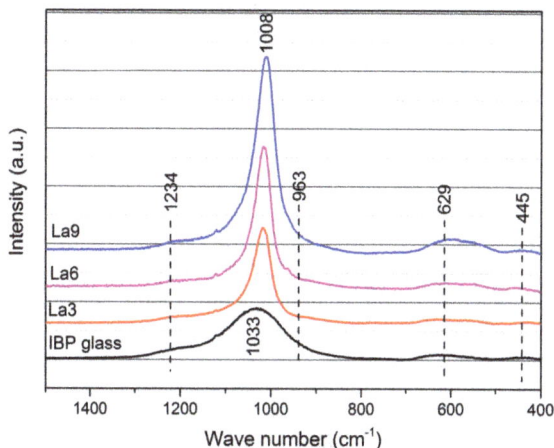

Fig. 8 *Ramna spectra of the La$_2$O$_3$ contained iron borophosphate glass samples in Series A.*

From Fig. 8, it can be seen that there are some modifications of the structure as the compositions of the studied samples change. With the addition of La$_2$O$_3$, the changes in Raman spectra are those: the peak ~ 629 cm^{-1} shifts towards lower wavenumbers; increases in intensity; ~ 1033 cm^{-1} shifts towards lower wavenumbers; increases in intensity; ~ 1234 cm^{-1} shifts towards lower wavenumbers; decreases in intensity. It is well known that glass structure can be influenced by glass modifying oxides that give rise to depolymerization of the glass network structure by converting bridging oxygen to non-bridging oxygen [40]. It seems that the main broad peak centered at 1033-1008 cm^{-1} comprises a number of overlapping components. The origin of the shape of this broad peak relates to a distribution of Q^0, Q^1 and Q^2 phosphate groups. In Raman spectrum, a shift towards lower wavenumbers may represent a more depolymerized phosphate network. Therefore, the addition of La_2O_3 to the base iron borophosphate glass results in a change in the distribution of phosphate structures defined by the Q nomenclature. The peaks both at ~ 1033 cm^{-1} and ~ 1255 cm^{-1} shifting to lower energies are ascribed to the conversion of Q^2 phosphate groups to Q^1 phosphate groups. Previous investigation reveals that the phosphate groups of iron phosphate glasses contain orthophosphate (Q^0 units), pyrophosphate (Q^1 units) and metaphosphate (Q^2 units). In the structural network of these glasses, the equilibrium between the pyrophosphate units and their disproportionation products according to the equation $2Q^1 \leftrightarrow Q^0 + Q^2$ may occur [17, 35]. Thus, the peak at ~ 1033 cm^{-1} increases in intensity, and the peak at ~ 1255 cm^{-1} decreases in intensity with increasing La_2O_3 content in the compositions of the studied IBP glasses.

Table 3 Assignments of Raman spectra for the studied samples.

Wavenumbers	Assignments	Ref.
~ 465 cm^{-1}	O–P–O bending modes of Q^0 $(PO_4)^{3-}$ units	[31, 32]
~ 575-629 cm^{-1}	Related to Q^1 $(P_2O_7)^{4-}$ units	[33]
~ 963 cm^{-1}	Asymmetric stretching of Q^0 $(PO_4)^{3-}$ monomer units	[32, 34]
~ 1033 cm^{-1}	Symmetric stretching of Q^1 $(P_2O_7)^{4-}$ dimer units	[35, 36]
~ 1234 cm^{-1}	Asymmetric stretching of Q^2 $(PO_3)^-$ metaphosphate units	[32, 34]

A band at ~ 630 cm^{-1} observed in Raman spectra for iron phosphate glasses [31, 35] has previously been assigned to Q^2 $(PO_3)^-$ metaphosphate units. Those authors claim that the band is originally indicative of Q^2 units resulting from disproportionation of Q^1 units to Q^0 and Q^2 units. Other investigations on modifier-containing phosphate glasses show a band centered ~ 690 cm^{-1} [32, 41] and this band has also been assigned to symmetric stretching of (P–O–P) bridging oxygens in Q^2 $(PO_3)^-$ metaphosphate units. The Raman shifts and band shapes varying with compositions for the band ~ 629 cm^{-1} in our spectra are different from those of both the ~ 690 cm^{-1} and ~ 630 cm^{-1} band. We do note that, with increase of La_2O_3 content, the Raman band at ~ 629-575 cm^{-1} proportionately increases in intensity with the band at ~ 1033 cm^{-1} (Fig. 7). Therefore we suggest that the 629 cm^{-1} band arises from dimer units, i.e. Q^1 $(P_2O_7)^{4-}$ units. The fact that both bands behave proportionately, suggesting that they have similar structural origins, has been discussed by other authors [32, 42]. There is also one publication discussing the presence of a Raman band at ~ 650 cm^{-1} and attributing it to Q^1 (P$_2$O$_7$)$^{4-}$ units [33] in the structure of modified phosphate glasses. Moreover, the Raman peaks between 525 and 583 cm^{-1} have been assigned to the overlapping vibrations involving iron oxygen polyhedral and (P$_2$O$_7$)$^{4-}$ units groups [43]. From the results of structure analysis, the reasons for the lower dissolution rate values (high chemical durability) of the studied iron borophosphate glasses are believed to be caused by the replacement of P–O–P bonds by the more hydration resistant M–O–P (M = Fe, La) bonds [6] and the predominate existence of hydration resistant Q^1 (P$_2$O$_7$)$^{4-}$ units in their structure.

4. Dissolution rate

Fig. 9(a) shows the aqueous dissolution rate (DR) values calculated from the weight loss of bulk samples immersed in 90 °C deionized water for 3, 7, 14, 21, 28 and 56 days. In general, the DR values of the samples are about 10^{-9} g·cm^{-2}·min^{-1} which are much lower than those of the commercial soda-lime-silica window glass and the iron borophosphate glass waste forms containing 30 wt.% simulated nuclear wastes measured in the same conditions [19]. Fig. 9(b) shows the detailed DR values of samples after immersing in 90 °C deionized water for 14 days. The figure shows that the DR values of the samples containing different amount of La$_2$O$_3$ are in the range of 2.89×10^{-9} g·cm^{-2}·min^{-1} to 4.96×10^{-9} g·cm^{-2}·min^{-1}. In Fig. 9(b), the borosilicate glass CVS-IS is a standard glass made by Pacific Northwest National Labs (PNNL) [6]. The nominal compositions of this glassy waste form can be found in the literature [6]. The short term DR values of the studied glasses are comparable to that of the standard borosilicate glass waste form and are higher than that of the common soda-lime glass [6, 19]. The obtained aqueous DR values indicate that the chemical durability of these glasses is comparable to widely used borosilicate glass waste forms and some base iron phosphate system glasses measured according to the same standard [6, 15, 44].

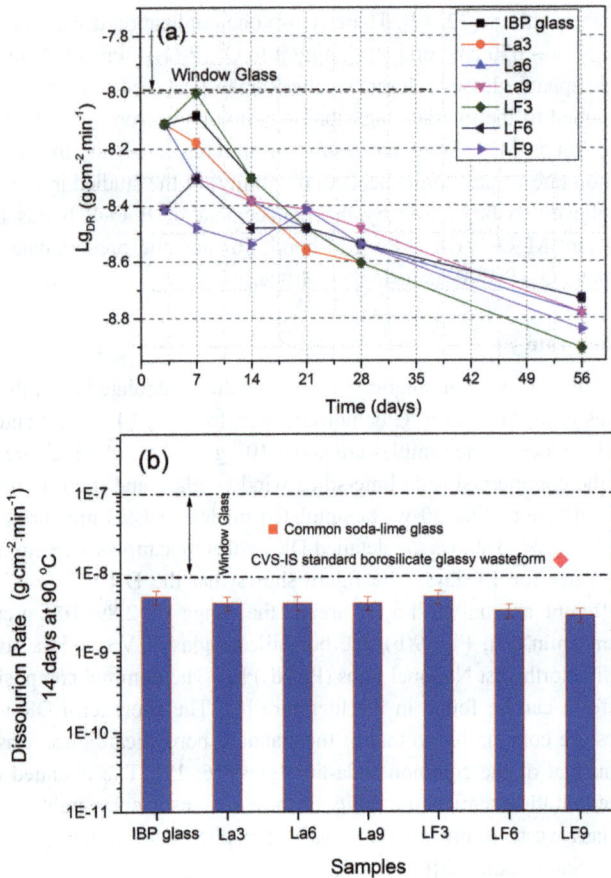

Fig. 9 *Dissolution rate of the samples immersed in 90 °C deionized water for different days. The DR value of window glass is between the two dashed lines at 1E-8 and 1E-7 in Fig. 9(b).*

Observation of any crystalline phases in a glass matrix is not desired since phase separation may degrade the chemical durability of waste glass. However, the samples containing more than 6 mol% La_2O_3 have monazite $LaPO_4$ crystalline phase. Fig. 9 shows that the chemical durability of all the samples, with and without monazite $LaPO_4$

crystalline phase, studied in the present study is high. This demonstrates that the formed monazite $LaPO_4$ crystalline phase insignificantly affect the DR of the base IBP glass. The reason for this result is that the formed monazite $LaPO_4$ crystalline phase itself possesses high aqueous resistance and good structure stability [45]. Several previous published literatures have also reported the similar phenomenon that the formation of crystalline clusters due to solubility limit is not affect the aqueous chemical durability of the base glass if the formed crystalline phases possess high aqueous resistance and good structure stability [15]. These results indicate that the studied iron borophosphate glasses containing La are potential hosts for the disposal of high-level nuclear wastes. Moreover, the acceptance of crystalline phase in a glass-based waste form will typically lead to higher waste loadings than those in a pristine glass waste form.

References

[1]   H. Darwish, Investigation of the durability of sodium calcium aluminum borosilicate glass containing different additives, Mater. Chem. Phys. 69 (2001) 36-44. http://dx.doi.org/10.1016/S0254-0584(00)00295-9

[2]   C.W. Kim, C.S. Ray, D. Zhu, D.E. Day, D. Gombert, A. Aloy, A. Moguš-Milankovic, M. Karabulut, Chemically durable iron phosphate glasses for vitrifying sodium bearing waste (SBW) using conventional and cold crucible induction melting (CCIM) techniques, J. Nucl. Mater. 322 (2003) 152-164. http://dx.doi.org/10.1016/S0022-3115(03)00325-8

[3]   H. Jena, B.K. Maji, R. Asuvathraman, K.V. Govindan Kutty, Effect of pyrochemical chloride waste loading on thermo-physical properties of borosilicate glass bonded Sr-chloroapatite composite, Mater. Chem. Phys. 162 (2015) 188-196. http://dx.doi.org/10.1016/j.matchemphys.2015.05.057

[4]   M.G. Mesko, D.E. Day, Immobilization of spent nuclear fuel in iron phosphate glass, J. Nucl. Mater. 273 (1999) 27-36. http://dx.doi.org/10.1016/S0022-3115(99)00020-3

[5]   S. Ibrahim, M.M. Morsi, Effect of increasing Fe_2O_3 content on the chemical durability and infrared spectra of $(25-x)Na_2O-xFe_2O_3-25PbO-50SiO_2$ glasses, Mater. Chem. Phys. 138 (2013) 628-632. http://dx.doi.org/10.1016/j.matchemphys.2012.12.030

[6]   D.E. Day, Z. Wu, C.S. Ray, et al, Chemically durable iron phosphate glass wasteforms, J. Non-Cryst. Solids 241 (1998) 1-12. http://dx.doi.org/10.1016/S0022-3093(98)00759-5

[7] K. Cheol-Woon, D.E. Day, Immobilization of Hanford LAW in iron phosphate glasses, J. Non-Cryst. Solids 331 (2003) 20-31. http://dx.doi.org/10.1016/j.jnoncrysol.2003.08.070

[8] K. Joseph, R. Asuvathraman, R.V. Krishnan, *et al*, Iron phosphate glass containing simulated fast reactor waste: Characterization and comparison with pristine iron phosphate glass, J. Nucl. Mater. 452 (2014) 273-280. http://dx.doi.org/10.1016/j.jnucmat.2014.05.038

[9] P.Y. Shih, Properties and FTIR spectra of lead phosphate glasses for nuclear waste immobilization, Mater. Chem. Phys. 80 (2003) 299-304. http://dx.doi.org/10.1016/S0254-0584(02)00516-3

[10] P.A. Bingham, R.J. Hand, S.D. Forder, Doping of iron phosphate glasses with Al_2O_3, SiO_2 or B_2O_3 for improved thermal stability, Mater. Res. Bull. 41 (2006) 1622-1630. http://dx.doi.org/10.1016/j.materresbull.2006.02.029

[11] F.H. ElBatal, Y.M. Hamdy, S.Y. Marzouk, Gamma ray interactions with V_2O_5-doped sodium phosphate glasses, Mater. Chem. Phys. 112 (2008) 991-100. http://dx.doi.org/10.1016/j.matchemphys.2008.07.005

[12] P.A. Bingham, R.J. Hand, S.D. Forder, A. Lavaysierre, F. Deloffre, S.H. Kilcoyne, I. Yasin, Structure and properties of iron borophosphate glasses, Phys. Chem. Glasses-B 47 (2006) 313-317.

[13] M. Karabulut, B. Yuce, O. Bozdogan, H. Ertap, G.M. Mammadov, Effect of boron addition on the structure and properties of iron phosphate glasses, J. Non-Cryst. Solids 357 (2011) 1455-1462. http://dx.doi.org/10.1016/j.jnoncrysol.2010.11.023

[14] P.A. Bingham, G. Yang, R.J. Hand, G. Möbus, Boron environments and irradiation stability of iron borophosphate glasses analysed by EELS, Solid State Sciences 10 (2008) 1194-1199. http://dx.doi.org/10.1016/j.solidstatesciences.2007.11.024

[15] M. Karabulut, C. Aydın, H. Ertap, M. Yüksek, Structure and properties of hafnium iron borophosphate glass-ceramics, J. Non-Cryst. Solids 411 (2015) 19-25. http://dx.doi.org/10.1016/j.jnoncrysol.2014.12.014

[16] F. Wang, Q.L. Liao, G.H. Xiang, S.Q. Pan, Thermal properties and FTIR spectra of K_2O/Na_2O iron borophosphate glasses, J. Mol. Struct. 1060 (2014) 176-181. http://dx.doi.org/10.1016/j.molstruc.2013.12.049

[17] F. Wang, Q.L. Liao, K.R. Chen, S.Q. Pan, M.W. Lu, Glass formation and FTIR spectra of CeO_2-doped $36Fe_2O_3$-$10B_2O_3$-$54P_2O_5$ glasses, J. Non-Cryst. Solids 409 (2015) 76-82. http://dx.doi.org/10.1016/j.jnoncrysol.2014.11.020

[18] F. Wang, Q.L. Liao, K.R. Chen, S.Q. Pan, M.W. Lu, The crystallization and FTIR spectra of ZrO_2-doped $36Fe_2O_3$-$10B_2O_3$-$54P_2O_5$ glasses and crystalline compounds, J. Alloys Compd. 611 (2014) 278-283. http://dx.doi.org/10.1016/j.jallcom.2014.05.117

[19] Q.L. Liao, F. Wang, K.R. Chen, S.Q. Pan, et al. FTIR spectra and properties of iron borophosphate glasses containing simulated nuclear wastes, J. Mol. Struct. 1092 (2015) 187-191. http://dx.doi.org/10.1016/j.molstruc.2015.03.034

[20] F. Wang, Q.L. Liao, Y.Y. Dai, H.Z. Zhu. Properties and vibrational spectra of iron borophosphate glasses/glass-ceramics containing lanthanum, Materials Chemistry and Physics, 2015, 166(C): 215-222. http://dx.doi.org/10.1016/j.matchemphys.2015.10.005

[21] M. Karabulut, E. Metwalli, R.K. Brow, Structure and properties of lanthanum–aluminum–phosphate glasses, J. Non-Cryst. Solids 283 (2001) 211-219. http://dx.doi.org/10.1016/S0022-3093(01)00420-3

[22] B. Qian, S. Yang, X. Liang, Y. Lai, L. Gao, G. Yin, Structural and thermal properties of La_2O_3-Fe_2O_3-P_2O_5 glasses, J. Mol. Struct. 1011 (2012) 153-157. http://dx.doi.org/10.1016/j.molstruc.2011.12.014

[23] D.A. Magdas, O. Cozar, V. Chis, I. Ardelean, The structural dual role of Fe_2O_3 in some lead-phosphate glasses, Vib. Spectrosc. 48 (2008) 251-254. http://dx.doi.org/10.1016/j.vibspec.2008.02.016

[24] P.Pascuta, G. Borodi, A. Popa, V. Dan, E. Culea, Influence of iron ions on the structural and magnetic properties of some zinc-phosphate glasses, Mater. Chem. Phys. 123 (2010) 767–771. http://dx.doi.org/10.1016/j.matchemphys.2010.05.056

[25] Y. Lai, X. Liang, G. Yin, S. Yang, J. Wang, H. Zhu, H. Yu, Infrared spectra of iron phosphate glasses with gadolinium oxide, J. Mol. Struct. 1004 (2011) 188-192. http://dx.doi.org/10.1016/j.molstruc.2011.08.003

[26] H.A. ElBatal, A.M. Abdelghany, I.S. Ali, Optical and FTIR studies of CuO-doped lead borate glasses and effect of gamma irradiation, J. Non-Cryst. Solids 358 (2012) 820-825. http://dx.doi.org/10.1016/j.jnoncrysol.2011.12.069

[27] M. Karabulut, M. Yüksek, G.K. Marasinghe, D.E. Day, Structural features of hafnium iron phosphate glasses, J. Non-Cryst. Solids 355 (2009) 1571-1573. http://dx.doi.org/10.1016/j.jnoncrysol.2009.06.005

[28] L. Baia, D. Muresan, M. Baia, J. Popp, S. Simon, Structural properties of silver nanoclusters–phosphate glass composites, Vib. Spectrosc. 43 (2007) 313-318. http://dx.doi.org/10.1016/j.vibspec.2006.03.006

[29] A. Majjane, A.Chahine, M.Et-tabirou, B. Echchahed, T. Do, P.M. Breen, X-ray photoelectron spectroscopy (XPS) and FTIR studies of vanadium barium phosphate glasses, Mater. Chem. Phys.143 (2014) 779-787. http://dx.doi.org/10.1016/j.matchemphys.2013.10.013

[30] L. Macalika, P.E. Tomaszewski, A. Matraszek, I. Szczygieł, P. Solarz, P. Godlewska, M. Sobczyk, J. Hanuza, Optical and structural characterisation of pure and Pr^{3+} doped $LaPO_4$ and $CePO_4$ nanocrystals, J. Alloys Compd. 509 (2011) 7458- 7465. http://dx.doi.org/10.1016/j.jallcom.2011.04.077

[31] A. Moguš-Milankovic, A. Šantic, S.T. Reis, K. Furic, D.E. Day, Mixed ion–polaron transport in $Na_2O–PbO–Fe_2O_3–P_2O_5$ glasses, J. Non-Cryst. Solids 342 (2004) 97-109. http://dx.doi.org/10.1016/j.jnoncrysol.2004.07.012

[32] P.A. Bingham, R.J. Hand, O.M. Hannant, S.D. Forder, S.H. Kilcoyne, Effects of modifier additions on the thermal properties, chemical durability, oxidation state and structure of iron phosphate glasses, J. Non-Cryst. Solids 355 (2009) 1526-1538. http://dx.doi.org/10.1016/j.jnoncrysol.2009.03.008

[33] G.K. Marasinghe, M. Karabulut, C.S. Ray, et al, Effects of nuclear waste components on redox equilibria, structural features, and crystallization characteristics of iron phosphate glasses, Ceram. Trans. 93 (2003) 195-201.

[34] G.S. Henderson, R.T. Amos, The structure of alkali germanophosphate glasses by Raman spectroscopy, J. Non-Cryst. Solids 328 (2003) 1-19. http://dx.doi.org/10.1016/S0022-3093(03)00478-2

[35] B. Qian, X. Liang, S. Yang, S. He, L. Gao, Effects of lanthanum addition on the structure and properties of iron phosphate glasses, J. Mol. Struct. 1027 (2012) 31-35. http://dx.doi.org/10.1016/j.molstruc.2012.05.078

[36] L. Zhang, R.K. Brow, A Raman Study of Iron–Phosphate Crystalline Compounds and Glasses, J. Am. Ceram. Soc. 94 (2011) 3123-3130. http://dx.doi.org/10.1111/j.1551-2916.2011.04486.x

[37] P. Mošner, M. Vorokht, L. Koudelka, L. Montagne, B. Revel, K. Sklepić, A. Moguš-Milanković, Effect of germanium oxide on the structure and properties of lithium borophosphate glasses, J. Non-Cryst. Solids 375 (2013) 1-6. http://dx.doi.org/10.1016/j.jnoncrysol.2013.05.009

[38] L. Koudelka, P. Mošner, M. Zeyer, C. Jäger, Structural study of PbO-B_2O_3-P_2O_5 glasses by NMR, Raman and infrared spectroscopy, Phys. Chem. Glasses 43 (2002) 102-107.

[39] J.W. Lim, M.L. Schmitt, R.K. Brow, S.W. Yung, Properties and structures of tin borophosphate glasses, J. Non-Cryst. Solids 356 (2010) 1379-1384. http://dx.doi.org/10.1016/j.jnoncrysol.2010.02.019

[40] M.W. Lu, F. Wang, K.R. Chen, Y.Y. Dai, Q. Liao, H.Z. Zhu, The crystallization and structure features of barium-iron phosphate glasses, Spectrochim. Acta Part A 148 (2015) 1–6. http://dx.doi.org/10.1016/j.saa.2015.03.121

[41] R.K. Brow, D.R. Tallant, W.L. Warren, A. McIntyre, D.E. Day, Spectroscopic studies of the structure of titanophosphate and calcium titanophosphate glasses, Phys. Chem. Glasses 38 (1997) 300-307.

[42] A. Mogus-Milankovic, K. Furic, D.E. Day, Scientific basis for nuclear waste management, Mat. Res. Soc. Symp. Proc. 663 (2001) 153-155. http://dx.doi.org/10.1557/PROC-663-153

[43] Y.M. Lai, X.F. Liang, S.Y. Yang, J.X. Wang, B.T. Zhang, Raman spectra study of iron phosphate glasses with sodium sulfate, J. Mol. Struct. 1013 (2012) 134-137. http://dx.doi.org/10.1016/j.molstruc.2012.01.025

[44] S.T. Reis, M. Karabulut, D.E. Day, Crystal chemistry of the monazite structure, J. Non-Cryst. Solids 292 (2001) 150-157. http://dx.doi.org/10.1016/S0022-3093(01)00880-8

[45] N. Clavier, R. Podor, N. Dacheux, Crystal chemistry of the monazite structure, J. Eur. Ceram. Soc. 31 (2011) 941–976. http://dx.doi.org/10.1016/j.jeurceramsoc.2010.12.019

CHAPTER 4

Crystallization studies of cerium containing iron borophosphate glasses/glass-ceramics

Fu Wang, Qilong Liao

School of Material Science and Engineering, Southwest University of Science and Technology, Mianyang 621010, PR China

Abstract

The properties of cerium containing iron borophosphate glasses/glass-ceramics are well implications for the properties of iron borophosphate based waste forms. To achieve high waste volume reduction, the content of nuclides usually exceeds their solibility limit in base glasses, which results in a crystalline phase formed in the corresponding waste forms. It is well known that if crystallite appears in a glassy phase, the crystallite will induce nucleation or provides surface to nucleate, resulting in much easier crystallization of the glassy material. Once crystallization occurs, the properties of the glassy waste form will dramatically deteriorate. In this chapter, the recent studies on the crystallization behaviors and crystallization kinetics of the cerium containing iron borophosphate glasses/glass-ceramics were summarized.

Keywords

Iron Borophosphate Glasses, Iron Borophosphate Based Glass-Ceramics, Cerium, Crystallization Behavior, Crystallization Kinetics

Contents

1.   Introduction

Vitrification is considered as one of the best methods for disposal of high-level radioactive wastes (HLWs) in industrial scale. Some HLWs waiting for disposal are rich in phosphates, sulfate, iron oxide, chromium oxide and heavy metals. These constituents are poorly soluble in borosilicate glasses which are widely used for HLWs immobilization. Thus, these constituents easily lead to phase separation in the structure of the resulting waste forms, and the chemical durability, structure stability and thermal stability of the waste forms dramatically deteriorate [1-3]. Therefore, iron phosphate glasses have received extensive attention in recent years, owing to their good ability to dissolve these problematic constituents [4, 5]. However, the radiation stability of iron phosphate waste forms is inferior to that of widely used borosilicate waste forms [6]. Bingham and Hand [7] studied the influence of B_2O_3 which has both high neutron absorption coefficient and mass absorption coefficient on the properties of iron phosphate glasses. Their obtained results show that doping of $40Fe_2O_3$-$60P_2O_5$ glass with 10 mol% B_2O_3 (iron borophosphate glasses) improves the thermal stability of the glass and insignificantly influences its chemical durability. Moreover, the radiation stability of the iron borophosphate glasses is comparable to borosilicate waste forms. Karabulut et al [8] studied doping modes of B_2O_3 on the properties of iron phosphate glasses. The results suggest that $36Fe_2O_3$-$10B_2O_3$-$54P_2O_5$ glass (IBP glass) possesses optimal properties. All these investigations show that iron borophosphate glasses are suitable for HLWs immobilization.

Cerium is widely used for simulating trivalent and tetravalent actinide radionuclides in study of HLWs immobilization [9-11]. Previous studies showed that the IBP based waste forms contain $CePO_4$ crystallite when the content of CeO_2 exceeds 9 mol% [12], and the formed $CePO_4$ crystallite improves the chemical durability of the waste forms [9]. Thus CeO_2 loading in the IBP base glasses increases. However, chemical durability is only one prerequisite property with which a waste form must to be satisfied. It is well known that glass phase is a metastable phase which can spontaneously crystallize to its crystalline state. This tendency is particularly obvious at high temperature. If crystallite appears in a glassy phase, the crystallite will induce nucleation or provides surface to nucleate, resulting in much easier crystallization of the glassy material. For up to hundreds of years of geological disposal, a glassy waste form unavoidably experiences high temperature in geological circumstance. Once crystallization occurs, the properties of the glassy waste form will dramatically deteriorate [2, 13]. Therefore, the resistance of glassy waste forms to crystallization at high temperature is significantly important. In this chapter, the effects of CeO_2 addition and the formed $CePO_4$ crystallite on the crystallization behaviors of the crystallized IBP glasses/glass-ceramics are discussed [14]. The discussed samples are marked by CeO_2 content, for instance, Ce3 sample represents $36Fe_2O_3–10B_2O_3–54P_2O_5$ glass doped with 3 mol% CeO_2. The real chemical compositions with respect to each constituent of as-prepared samples are listed in Table 1.

Table 1 *The chemical compositions of the as-prepared samples.*

Samples	Compositions (mol %)			
	Fe_2O_3	B_2O_3	P_2O_5	CeO_2
Ce0	36.00	10.10	53.90	0
Ce3	34.92	9.71	52.35	3.04
Ce6	33.84	9.42	50.74	5.96
Ce9	32.78	9.16	49.16	9.12
Ce12	31.70	8.84	47.55	12.09
Ce15	30.62	8.55	45.96	15.13
Ce18	29.10	8.08	44.18	18.34

2.   Thermal analysis

2.1   Glass forming ability

The identified crystalline phases in the prepared samples are shown in Fig. 1 and the JCPDS standard cards for $CePO_4$ (PDF# 32-0199) and $FePO_4$ crystals (PDF# 29-0715) are also shown in the figure for analysis assistance. The as-cast and annealed IBP glass with less than 9 mol% of CeO_2 is fully amorphous (Fig. 1 (a)). Comparing with the JCPDS standard cards, It is known that when doped CeO_2 content is more than 9 mol% (including 9 mol%), monazite $CePO_4$ crystal phase (PDF *No.* 32-0199, space group: Monoclinic, *P21/n*) appears in the structure of the studied compounds. Furthermore, with continual increase in doped-CeO_2, diffraction peaks represented to $CePO_4$ crystal grow strong, which may due to strong ability of Ce^{3+} to orderly arrange surrounding atoms or ions according to its own required coordination numbers. When doped CeO_2 reaches 18 mol% (for Ce18 sample), another crystal phase, $FePO_4$ (PDF *No.* 29-0715, space group:

Hexagonal, $P321$) is detected (Fig. 1 (b)), demonstrating that Fe is crystallized from the base glass. This is because, for one thing, CeO_2 will increase the disorder structure of the glass to satisfy the high coordination number of Ce element, which decreases the structural stability of the IBP glass. For another, the appeared $CePO_4$ crystal is easy to induce $FePO_4$ crystallized from the base glass.

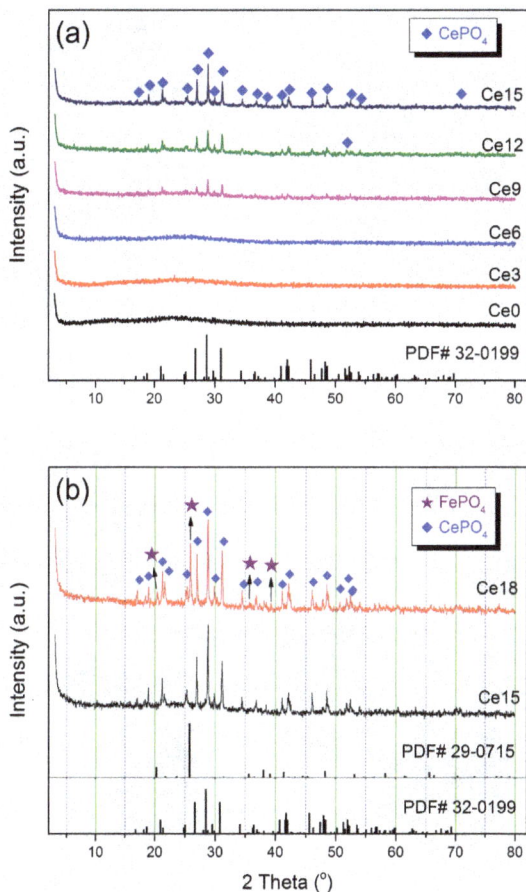

Fig. 1 *(a) XRD patterns of the partial prepared samples and (b) attentive comparison of XRD patterns between Ce15 and Ce18 sample, PDF# 32-0199 and PDF# 29-0715 are JCPDS standard cards for $CePO_4$ and $FePO_4$ crystals respectively.*

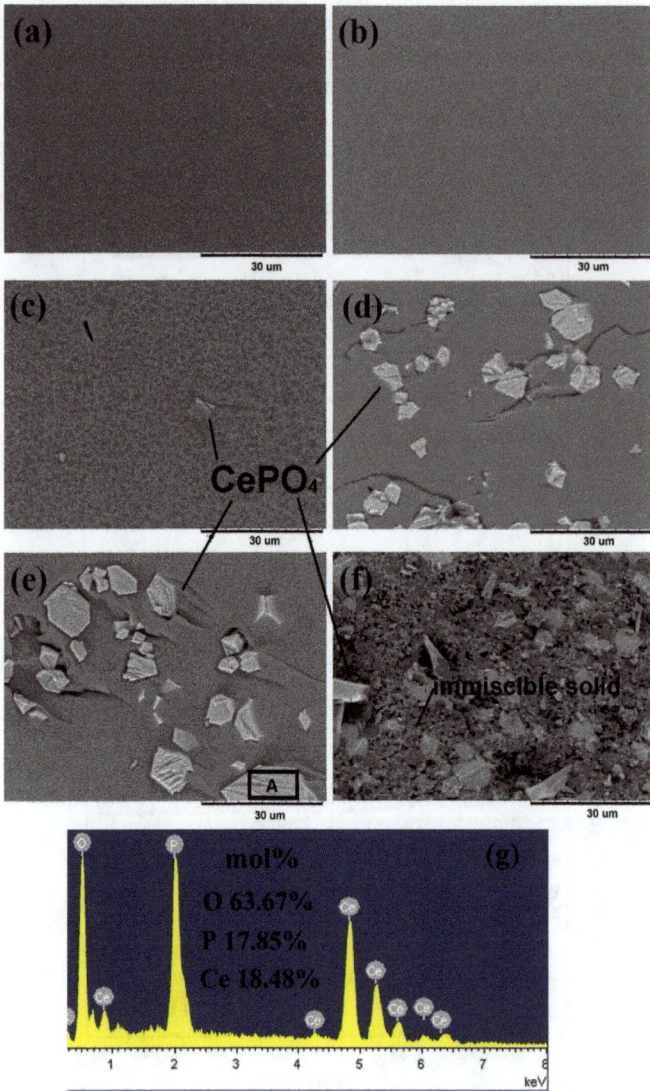

Fig. 2 SEM of the (a) Ce0, (b) Ce6, (c) Ce9, (d) Ce12 (e) Ce15 and (e) Ce18 sample and (g) EDS pattern of the crystal in the A areas of the SEM figure.

The results of SEM images and EDS pattern of the appeared crystal are consistent to that of photos and XRD analysis, as shown in Fig. 2. When the content of the doped CeO_2 is up to 9 mol%, the studied IBP glasses show phase separation (crystal appears) (Figs. 2(c)-(e)), although they still have extremely compact structure, thus they still have smooth surface. The EDS pattern show that the constituent elements of the appeared crystal are Ce, P and O, and the molar ratios of Ce:P:O are about 1.03:1:3.65 (Fig. 2(g)) which is close to the value 1:1:4. By combining the XRD analysis results, $CePO_4$ crystal is determined. Combining with the photos, SEM images, EDS and XRD patterns analysis, it can be concluded that the studied IBP glass can immobilize 15 mol% of CeO_2 at least, merely monazite $CePO_4$ forms for the IBP glass doped with CeO_2 between 9 and 15 mol%. The simplified glass formation region as a function of CeO_2 content for samples melted at 1200 °C are shown in Fig. 3.

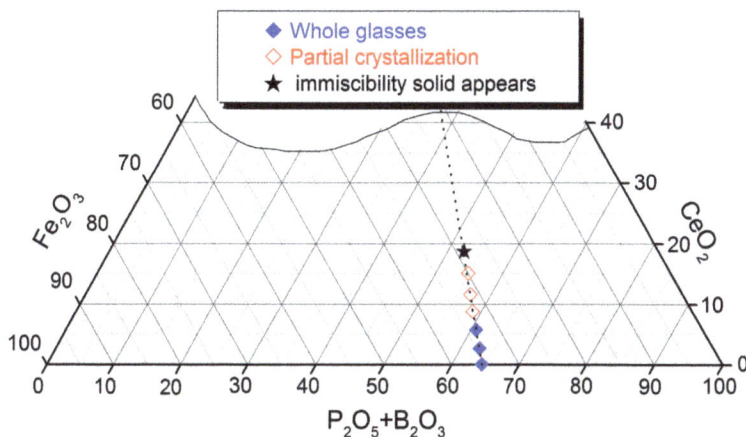

Fig. 3 Simplified cerium iron borophosphate glasses formation diagram.

2.2 Thermal analysis and kinetics of the cerium containing iron borophosphate glasses/glass-ceramics

DTA curves of typical samples are shown in Fig. 4. It is observed that all the patterns are similar in nature and characterized with an endothermic peak. Such peak represents glass transition phenomenon and the onset temperature of endothermic peak corresponds to glass transition temperature (T_g). A shifting of glass transition temperatures to higher values for Ce6 sample with 6 mol% CeO_2 (Table 2) indicates an increase in strengthening of structure [15]. The increase in T_g can be attributed to significantly larger ion radius of Ce^{3+}/Ce^{4+} which exists in glass structure than that of Fe^{3+}/Fe^{4+}, which leads to more difficult structure modification. However, From Ce6 sample to Ce9 sample which contain higher CeO_2, a shifting of T_g to lower values occurs because $CePO_4$ crystalline appears for Ce9 sample. The DTA curves also show that doping with CeO_2 could decrease the crystallization onset temperature (T_r), seeing in the corresponding onset temperature to the first exothermic peak, and crystallization peak temperature (T_p) (Table 2), because Ce element makes the studied IBP glasses crystallize more easily due to its high coordination number, free oxygen supplier as glass modifying oxides and depolymerization in glass structure. Combining with induction of $CePO_4$ crystalline, T_r and T_p decrease more sharply (sample Ce9). In addition, Evaluations of thermal stability have been discussed through Hruby's method [15-17]. The greater temperature difference between T_r and T_g, the more stable the glass is. Namely, an increase of (T_r-T_g) value indicates an increase in thermal stability [17]. For above-mentioned reasons, the (T_r-T_g) values also decrease (Table 2), indicating that doping with CeO_2 also decreases thermal stability of the studied IBP glasses.

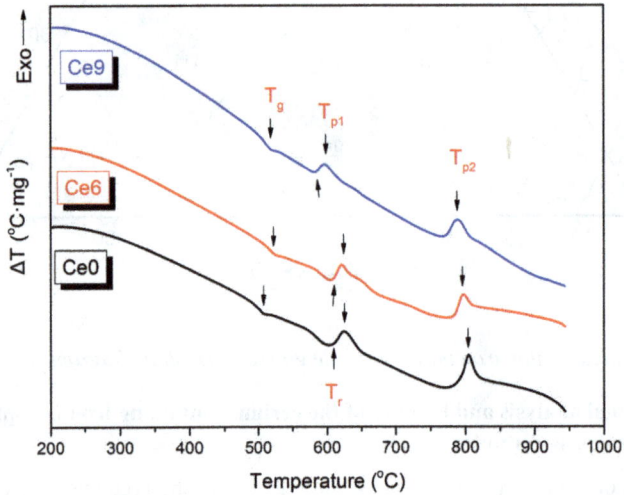

Fig. 4 *DTA curves of typical samples.*

Table 2 DTA parameters of typical samples.

Samples	Ce0	Ce6	Ce9
$T_g \pm 1$ (℃)	510.7	525.6	520.2
$T_r \pm 1$ (℃)	606.0	608.8	583.7
$T_{p1} \pm 1$ (℃)	625.4	621.1	596.4
$T_{p2} \pm 1$ (℃)	805.5	798.2	789.9
T_r-$T_g \pm 1$ (℃)	95.3	83.2	63.5

From the heating rate dependence of glass transition temperature (T_g) and crystallization peak temperature (T_p), the Avrami exponent (n), apparent activation energies for the crystallization and glass transition of the prepared samples have been calculated using Kissinger method [18], as shown in Table 3. The crystallization activation energies for the first crystallization peak (E_c for P1) and the second crystallization peak (E_c for P2) of the IBP glass are 211.77 KJ·mol^{-1} and 564.31 KJ·mol^{-1} respectively. And CeO$_2$ doping insignificantly affects the E_c for P1 but increases the E_c for P2. Moreover, the preexisting CePO$_4$ crystallite increases the E_c for P1 and the activation energies of glass transition peak (E_g). However, the CePO$_4$ crystallite obviously decreases the E_c for P2.

Table 3 The values of activation energies for T_g, the first and second crystallization temperature peaks.

Samples	E_g (KJ·mol^{-1})	E_c (KJ·mol^{-1})	
		P1	P2
Ce0	710.89 ± 1.15	211.77 ± 0.53	564.31 ± 0.87
Ce6	673.81 ± 2.76	210.11 ± 0.18	574.86 ± 0.20
Ce9	804.59 ± 1.73	232.39 ± 0.11	450.98 ± 0.28

3. Crystalline behaviors

3.1 XRD analysis

According to the measured DTA curves, two crystallization temperature peaks are observed at 590 - 625 °C and 785 - 805 °C respectively. Combining the heating rate of DTA measurement, the heat-treatment temperatures are determined to be 650 °C and 850

°C to investigate the crystalline behaviors of the prepared glasses/glass-ceramics. Fig. 5(a) shows the XRD pattern of the IBP glasses/glass-ceramics annealed at 650 °C and Fig. 5(b) is their magnifying part in the 2θ range of 20° to 35°. The figure shows that the phases of the IBP glass annealed at 650 °C for 10h are $Fe_4(PO_4)_2O$ (PDF *No.* 74-1443) and $Fe_2(PO_4)O$ (PDF *No.* 85-2386). The intensity of the characteristic peaks representing the above-mentioned crystalline phases decrease with increasing CeO_2 content, which indicates that CeO_2 addition suppresses the crystallization of the IBP glasses/glass-ceramics. When the content of CeO_2 is 9 mol%, some peaks representing $CePO_4$ crystal (PDF *No.* 32-0199) are observed. However, this does not indicate that $CePO_4$ phase appears in the crystallized IBP glasses/glass-ceramics. The detected $CePO_4$ crystal is the pre-existing $CePO_4$ crystallite according to our previous study [12]. From Fig. 5 and Table 4, it can be concluded that the pre-existing $CePO_4$ crystallite significantly suppresses the crystallization of the IBP glasses/glass-ceramics, which leads to the decreasing content of $Fe_4(PO_4)_2O$ and $Fe_2(PO_4)O$. Moreover, a small amount of another crystal phase, $FePO_4$ (PDF *No.* 84-0876), is detected when the $CePO_4$ crystallite pre-exists in the unannealed IBP glass-ceramics.

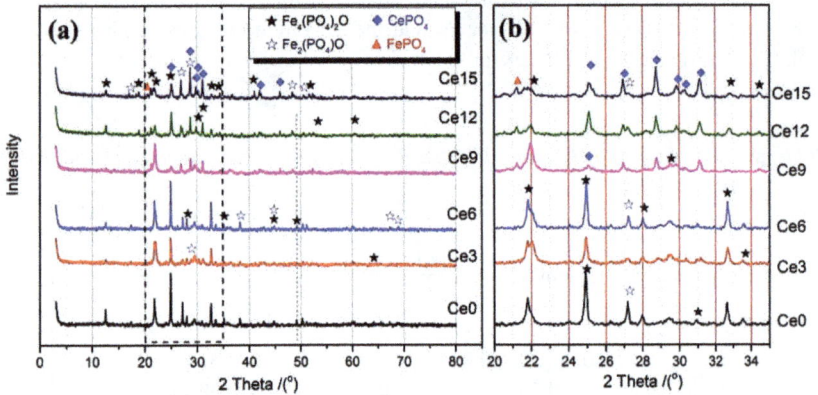

Fig. 5 *XRD patterns of the IBP glasses/glass-ceramics annealed at 650 °C.*

Table 4 *The phases and their approximate content of the IBP glasses/glass-ceramics annealed at 650 °C.*

Samples	Phases (Approximate content)
Ce0	$Fe_2(PO_4)O$ (main), $Fe_4(PO_4)_2O$ (main)
Ce3	$Fe_2(PO_4)O$ (main), $Fe_4(PO_4)_2O$ (main)
Ce6	$Fe_2(PO_4)O$ (limit), $Fe_4(PO_4)_2O$ (main)
Ce9	$Fe_2(PO_4)O$ (limit), $Fe_4(PO_4)_2O$ (limit), $CePO_4$(trace), $FePO_4$ (very trace)
Ce12	$Fe_2(PO_4)O$ (trace), $Fe_4(PO_4)_2O$ (limit), $CePO_4$(main), $FePO_4$ (very trace)
Ce15	$Fe_2(PO_4)O$ (trace), $Fe_4(PO_4)_2O$ (limit), $CePO_4$(main), $FePO_4$ (very trace)

Fig. 6(a) shows the XRD pattern of the IBP glasses/glass-ceramics annealed at 850 °C and Fig. 6(b) is their magnifying part in the 2θ range of 20° to 35°. The figure shows that the phases of the IBP glass annealed at 850 °C for 10h are $Fe_4(PO_4)_2O$, $Fe_2(PO_4)O$ and a small amount of $FePO_4$. CeO_2 addition also suppresses the crystallization of the IBP glasses/glass-ceramics when the selected annealing temperature is 850 °C. When the $CePO_4$ crystallite pre-exists, the content of $Fe_4(PO_4)_2O$ and $Fe_2(PO_4)O$ in the crystallized IBP glass-ceramics significantly decreases (Table 5). Moreover, some characteristic peaks representing $CePO_4$ crystal are observed for Ce6 sample, indicating that $CePO_4$ phase appears in the Ce6 sample when the annealing temperature is 850 °C (According to previous study, no $CePO_4$ crystallite is detected in the IBP glass containing 6 mol% CeO_2 [12]). It is known that the content of the pre-existing $CePO_4$ crystallite increases with increasing CeO_2 content [12], but the increasing content of the pre-existing $CePO_4$ crystallite does not lead to significant increase in content of $Fe_4(PO_4)_2O$ and $Fe_2(PO_4)O$ (comparing the XRD curves of Ce9, Ce12 and Ce15). This result indicates that the relationship between the content of $Fe_4(PO_4)_2O$ and $Fe_2(PO_4)O$ and the increasing quantity of the pre-existing $CePO_4$ crystallite is negligible. A quite small amount of the pre-existing $CePO_4$ crystallite can effectively suppress the crystallization of the IBP glasses/glass-ceramics. It is worth noting that the peaks representing B-containing phases are not observed for all IBP glasses/glass-ceramics annealed at both 650 °C and 850 °C, because glass forming ability of B_2O_3 is stronger than that of P_2O_5 [19].

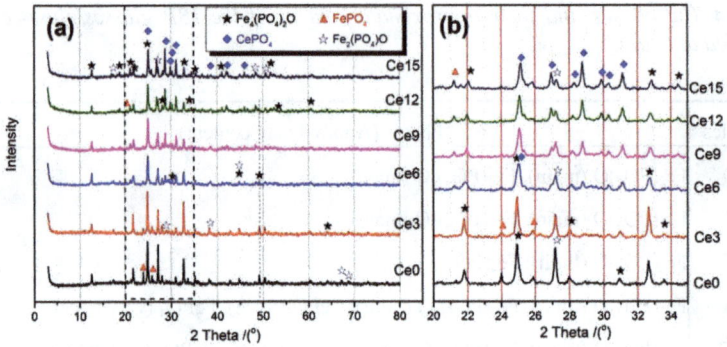

Fig. 6 *XRD patterns of the IBP glasses/glass-ceramics annealed at 850 °C.*

Table 5 *The phases and their approximate content of the IBP glasses/glass-ceramics annealed at 850 °C.*

Samples	Phases (Approximate content)
Ce0	$Fe_2(PO_4)O$ (main), $Fe_4(PO_4)_2O$ (main), $FePO_4$ (trace)
Ce3	$Fe_2(PO_4)O$ (main), $Fe_4(PO_4)_2O$ (main), $FePO_4$ (trace)
Ce6	$Fe_2(PO_4)O$ (limit), $Fe_4(PO_4)_2O$ (main), $FePO_4$ (trace), $CePO_4$(trace),
Ce9	$Fe_2(PO_4)O$ (limit), $Fe_4(PO_4)_2O$ (limit), $FePO_4$ (trace), $CePO_4$(limit)
Ce12	$Fe_2(PO_4)O$ (trace), $Fe_4(PO_4)_2O$ (limit), $FePO_4$ (very trace), $CePO_4$(main)
Ce15	$Fe_2(PO_4)O$ (trace), $Fe_4(PO_4)_2O$ (limit), $FePO_4$ (very trace), $CePO_4$(main)

3.2 SEM and EDS analysis

Fig. 7 shows the SEM of the IBP glasses/glass-ceramics annealed at 650 °C. The figure demonstrates that the microstructure of the samples is compact, because the main phases in the IBP glasses/glass-ceramics annealed at 650 °C are $Fe_4(PO_4)_2O$ and $Fe_2(PO_4)O$, and the compactness difference of the crystals and the base glass is small [19]. For the IBP glasses/glass-ceramics annealed at 850 °C, the microstructure of Ce0 and Ce3 sample is still compact, but the Ce6, Ce9, Ce12 and Ce15 sample show loose microstructure. Many pores are observed in their microstructure, as shown in Fig 8. According to XRD analysis, these samples contain a high content of $CePO_4$ crystal, while the content of $Fe_4(PO_4)_2O$ and $Fe_2(PO_4)O$ crystals is low. Moreover, the structure of $CePO_4$ crystal is very dense. Thus, the microstructure of these samples becomes loose due to the high content of dense $CePO_4$ crystal. The EDS pattern of the gray grains in SEM pictures, as shown in Fig. 9, shows that the constituents of these grains are Ce, P and O element and the average molar ratios of Ce : P : O are 17.66 : 16.88: 65.46 (about 1 : 1 : 4). Combing with the XRD analysis, the phase of these gray grains is determined to also be $CePO_4$. Previous study indicates that these grains are the formed $CePO_4$ crystallite in the IBP glass-ceramics [12]. Thus, the crystallized IBP glasses/glass-ceramics contain no or very limited $CePO_4$ phase when the annealing temperature is 650 °C, so the microstructure of the IBP glasses/glass-ceramics annealed at 650 °C is still compact (Fig. 7).

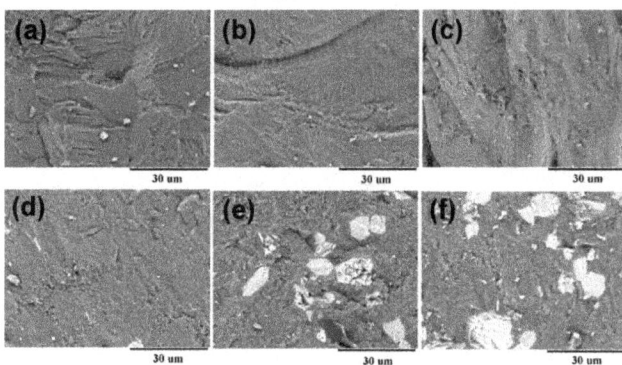

Fig. 7 *SEM of (a) Ce0, (b) Ce3, (c) Ce6, (d) Ce9, (e) Ce12 and (f) Ce15 sample (annealing at 650 °C).*

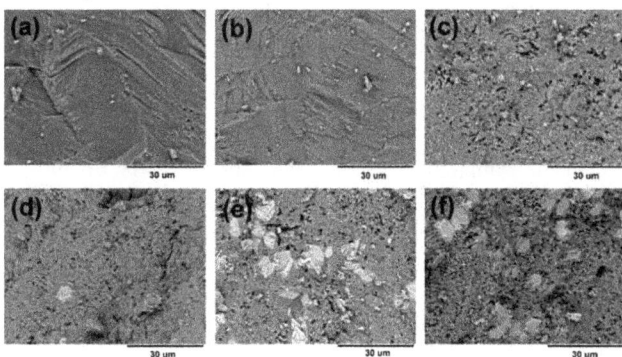

Fig. 8 *SEM of (a) Ce0, (b) Ce3, (c) Ce6, (d) Ce9, (e) Ce12 and (f) Ce15 sample (annealing at 850 °C).*

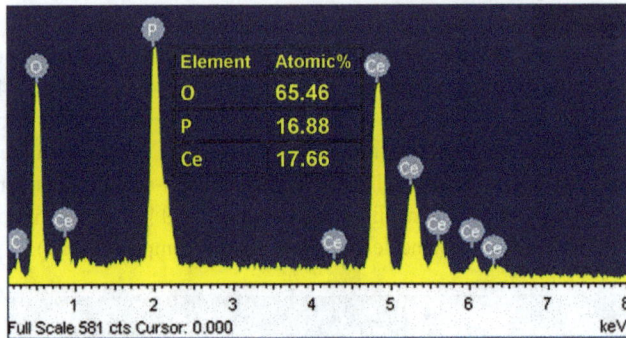

Element	Atomic%
O	65.46
P	16.88
Ce	17.66

Full Scale 581 cts Cursor: 0.000   keV

Fig. 9 *EDS pattern of the grey crystal in SEM pictures.*

3.3   Structure analysis

The transformation of the phases for the same sample annealed at the two selected temperatures cannot give rise to significant variation of infrared spectra. Therefore, in order to investigate the effect of CeO_2 addition on the structure of the samples, infrared spectra of the IBP glasses/glass-ceramics annealed at 850 °C are measured and shown in Fig. 10. In the figure, the absorption band at 420 cm^{-1} is associated with the vibration of $[FeO_6]$ group [20], the absorption bands at 450 cm^{-1} and 490 cm^{-1} are caused by the vibration of $[CeO_4]$ and $[FeO_4]$ group respectively [21, 22]; The band at about 520 cm^{-1} is attributed to the bending vibration of P–O–P in $(PO_4)^{3-}$ group [23]; The absorption bands at 580 cm^{-1} and 599 cm^{-1} principally are assigned to the bending vibration of O–P–O in pyrophosphate Q^1 $(P_2O_7)^{4-}$ group [24]; The band at 636 cm^{-1} corresponds to the stretching vibration of Fe–O–P bond [23]; The band at about 895 cm^{-1} is attributed to the stretching vibration of B-O bond in $[BO_4]$ group [25]. Moreover, the small band at 960 cm^{-1} has been attributed to the symmetric stretching vibration of O–B–O bond in $[BO_4]$ group [26]. The broad band between 995 ~ 1030 cm^{-1} is attributed to the antisymmetric stretching vibration of $[PO_4]$ tetrahedron [27], and the band at 1080 cm^{-1} is attributed to the symmetric stretching vibration of PO_4^{3-} tetrahedron (PO^- ionic group) [27]; The band at about 1230 cm^{-1} is attributed to the stretching vibration of Fe–O–P bond [27]; The absorption band attributed to the vibration of $[BO_3]$ group is located at about 1330 cm^{-1} [24]. In this study, the absorption bands at 1384 cm^{-1} and 1450 cm^{-1} are mainly assigned to the stretching vibration of P=O bond [27].

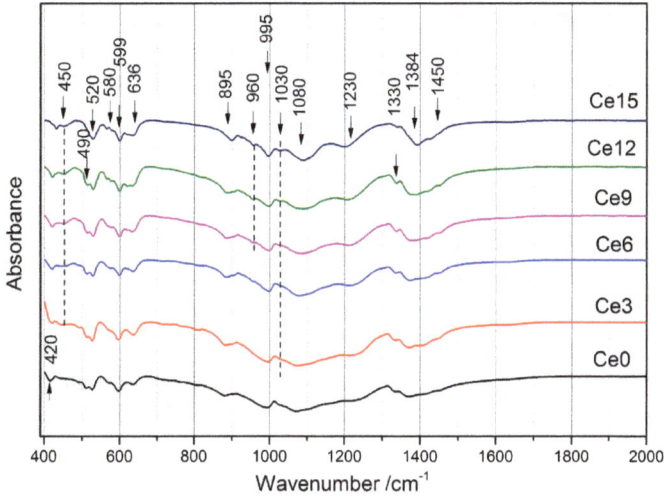

Fig. 10 *Infrared spectra of the crystallized IBP glasses/glass-ceramics (annealing at 850 °C).*

It can be seen from Fig. 10 that the main structural units of the samples are [PO$_4$] and [BO$_4$] tetrahedron. Moreover, it is worth noting that the structural units of the CeO$_2$-doped IBP glasses/glass-ceramics before heat-treatment do not contain [CeO$_4$] unit [12], but after heat-treatment, the absorption band at 450 cm^{-1} which has been assigned to the vibration of [CeO$_4$] unit is observed. This result indicates that [CeO$_4$] unit exists in the structure of the crystallized CeO$_2$-doped IBP glasses/glass-ceramics. The content of Fe$_4$(PO$_4$)$_2$O and Fe$_2$(PO$_4$)O phases decreases with increasing CeO$_2$ content in samples because CeO$_2$ addition suppresses the crystallization of the IBP glasses/glass-ceramics, which results in the decreased content of [FeO$_4$] group, thus the intensity of the band at about 490 cm^{-1} decreases. Particularly, the pre-existing CePO$_4$ crystallite shapely makes the content of Fe$_4$(PO$_4$)$_2$O and Fe$_2$(PO$_4$)O decrease, the absorption band near 490 cm^{-1} becomes much weaker. The intensity of the absorption bands at 636 cm^{-1} and 1230 cm^{-1} increase with the increasing content of CeO$_2$ and the band at 1230 cm^{-1} also shifts to low wave number, indicating that the bond length of Fe–O–P bond becomes longer and the content of Fe–O–P bond increases. These phenomenon are also mainly attributed to the decreased content of the Fe$_4$(PO$_4$)$_2$O and Fe$_2$(PO$_4$)O in samples. Meanwhile, due to the same reason and the introduction of non-bridge oxygen by CeO$_2$ addition [12], the content of [BO$_4$] group increases and that of [BO$_3$] group decreases, thus the bands at 895

cm^{-1} and at 960 cm^{-1} increase and the band at 1330 cm^{-1} decreases in intensity respectively. Although CeO_2 addition suppresses the crystallization of the IBP glasses/glass-ceramics, it induces the formation of $CePO_4$ phase and a small amount of $FePO_4$. Furthermore, the structural arrangement of $[PO_4]$ in $FePO_4$ and $CePO_4$ crystal is better ordered comparing with $Fe_4(PO_4)_2O$ and $Fe_2(PO_4)O$ crystal [19], and their bond strength is stronger. Therefore, the bands at 520 cm^{-1} and 995 ~ 1000 cm^{-1}, which have been assigned to the vibration of $[PO_4]$ tetrahedron, increase in intensity and shift to high wave number. This is also why the bands at 1384 cm^{-1} and 1450 cm^{-1} increase in intensity.

References

[1]   D.E. Day, Z. Wu, C.S. Ray. Chemically durable iron phosphate glass wasteforms. J. Non-Cryst. Solids 241 (1998) 1–12. http://dx.doi.org/10.1016/S0022-3093(98)00759-5kl

[2]   S. Schuller, O. Pinet, A. Grandjean, et al. Phase separation and crystallization of borosilicate glass enriched in MoO_3, P_2O_5, ZrO_2, CaO. J. Non-Cryst. Solids 354 (2008) 296–300. http://dx.doi.org/10.1016/j.jnoncrysol.2007.07.041

[3]   R.K. Mishra, R. Mishra, C.P. Kaushik, et al. Effect of ThO_2 on ionic transport behavior of barium borosilicate glasses. J. Nucl. Mater. 392 (2009) 1–5. http://dx.doi.org/10.1016/j.jnucmat.2009.03.001

[4]   F. Wang, Q.L Liao, Y.Y Dai, H.Z Zhu. Properties and vibrational spectra of iron borophosphate glasses/glass-ceramics containing lanthanum, Mater. Chem. Phys. 166 (2015) 215-222. http://dx.doi.org/10.1016/j.matchemphys.2015.10.005

[5]   P. Sengupta. A review on immobilization of phosphate containing high level nuclear wastes within glass matrix–Present status and future challenges. J. Hazard. Mater. 235 (2012) 17–28. http://dx.doi.org/10.1016/j.jhazmat.2012.07.039

[6]   P.A. Bingham, G. Yang, R.J. Hand, et al. Boron environments and irradiation stability of iron borophosphate glasses analysed by EELS. Solid State Sci. 10 (2008) 1194–1199. http://dx.doi.org/10.1016/j.solidstatesciences.2007.11.024

[7]   P.A. Bingham, R.J. Hand, S.D. Forder, et al. Structure and properties of iron borophosphate glasses. Phys. Chem. Glasses-B 47 (2006) 313–317.

[8]   M. Karabulut, B. Yuce, O. Bozdogan, et al. Effect of boron addition on the structure and properties of iron phosphate glasses. J. Non-Cryst Solids 57 (2013) 1455–1462.

[9]   Q.L. Liao, F. Wang, S.Q. Pan, et al. Structure and chemical durability of Ce-doped iron borophosphate glasses. J. Nucl. Radiochem. 32 (2010) 336–341.

[10]   P. Li, X.G. Ding, H. Yang, et al. Effect of mass fraction of CeO_2 on leaching behavior of aluminoborosilicate glass. Chinese J. Inorg. Chem. 29 (2013) 709–714.

[11]   Q.L. Liao, K.R. Chen, F. Wang, et al. The structure and performances of iron borophosphate glass wasteforms with simulated high sodium high level radioactive waste. J. Chin. Cera. Soc. 42 (2014) 119–124.

[12]   F. Wang, Q.L. Liao, K.R. Chen, et al. Glass formation and FTIR spectra of CeO_2-doped $36Fe_2O_3$-$10B_2O_3$-$54P_2O_5$ glasses. J. Non-Cryst. Solids 409 (2015) 76–82. http://dx.doi.org/10.1016/j.jnoncrysol.2014.11.020

[13]   G.D. Marsily. High level nuclear waste isolation: borosilicate glass versus crystals. Nature 278 (1979) 210–212. http://dx.doi.org/10.1038/278210a0

[14]   F. Wang, Q.L. Liao, H.Z. Zhu, Y.Y. Dai, H. Wang. Crystallization of cerium containing iron borophosphate glasses/glass-ceramics and their spectral properties, Journal of Molecular Structure, 2016, 1109: 226-231. http://dx.doi.org/10.1016/j.molstruc.2016.01.017

[15]   F. Wang, Q.L. Liao, G.H. Xiang, S.Q. Pan, Thermal properties and FTIR spectra of K_2O/Na_2O iron borophosphate glasses. J. Mol. Struct. 1060 (2014) 176-181. http://dx.doi.org/10.1016/j.molstruc.2013.12.049

[16]   P.A. Bingham, R.J. Hand, S.D. Forder, Doping of iron phosphate glasses with Al_2O_3, SiO_2 or B_2O_3 for improved thermal stability, Mater. Res. Bull. 41 (2006) 1622-1630. http://dx.doi.org/10.1016/j.materresbull.2006.02.029

[17]   A. Hrubý, Evaluation of glass-forming tendency by means of DTA, Czechoslovak Journal of Physics 22 (1972) 1187-1193. http://dx.doi.org/10.1007/BF01690134

[18]   F. Wang, Q.L. Liao, H.Z. Zhu, Y.Y. Dai, H. Wang, Crystallization kinetics and glass transition kinetics of iron borophosphate glass and CeO_2-doped iron borophosphate compounds, J. Alloys Compd. (2016). http://dx.doi.org/10.1016/j.jallcom.2016.06.066

[19]   F. Wang, Q.L. Liao, K.R. Chen, et al. The crystallization and FTIR spectra of ZrO_2-doped $36Fe_2O_3$-$10B_2O_3$-$54P_2O_5$ glasses and crystalline compounds. J. Alloys Compd. 611 (2014) 278–283. http://dx.doi.org/10.1016/j.jallcom.2014.05.117

[20]   N. Sdiri, H. Elhouichet, B. Azeza, et al. Studies of $(90-x)P_2O_5-xB_2O_3-10Fe_2O_3$ glasses by Mossbauer effect and impedance spectroscopy methods. J. Non-Cryst Solids 371 (2013) 22–27. http://dx.doi.org/10.1016/j.jnoncrysol.2013.04.002

[21]   G.P. Singh, P. Kaur, S. Kaur, et al. Conversion of Ce^{3+} to Ce^{4+} ions after gamma ray irradiation on $CeO_2-PbO-B_2O_3$ glasses. Physica B 408 (2013) 115–118. http://dx.doi.org/10.1016/j.physb.2012.09.005

[22]   C. Dayanand, G. Bhikshamaiah, V.J. Tyagaraju, et al. Structural investigations of phosphate glasses: a detailed infrared study of the $x(PbO)-(1-x) P_2O_5$ vitreous system. J. Mater. Sci. 31 (1996) 1945–1967. http://dx.doi.org/10.1007/BF00356615

[23]   B. Qian, S.Y. Yang, X.F. Liang, et al. Structural and thermal properties of $La_2O_3-Fe_2O_3-P_2O_5$ glasses. J. Mol. Struct. 1011 (2012) 153–157. http://dx.doi.org/10.1016/j.molstruc.2011.12.014

[24]   A.M. Abdelghany, M.A. Ouis, M.A. Azooz, et al. Defect formation of gamma irradiated MoO_3-doped borophosphate glasses. Spectrochim. Acta A 114 (2013) 569–574. http://dx.doi.org/10.1016/j.saa.2013.05.023

[25]   F.H. Elbatal, S. Ibrahim, A.M. Abdelghany. Optical and FTIR spectra of NdF3-doped borophosphate glasses and effect of gamma irradiation. J. Mol. Struct. 1030 (2012) 107–112. http://dx.doi.org/10.1016/j.molstruc.2012.02.049

[26]   K. Srinivasulu, I. Omkaram, H. Obeid, et al. Structural and Magnetic properties of Gd^{3+} ions in Sodium-lead borophosphate glasses. J. Mol. Struct. 1036 (2013) 63–70. http://dx.doi.org/10.1016/j.molstruc.2012.09.041

[27]   D.A. Magdas, O. Cozar, V. Chis, et al. The structural dual role of Fe_2O_3 in some lead-phosphate glasses. Vib. Spectrosc. 48 (2008) 251–254. http://dx.doi.org/10.1016/j.vibspec.2008.02.016

CHAPTER 5

Spectroscopic properties and energy transfer parameters of Nd^{3+} and Sm^{3+} doped lithium borate glasses

J. Anjaiah[1,2], C. Laxmikanth[1]

[1] Department of Physics, The University of Dodoma, Tanzania, East Africa.

[2] Department of Physics, Geethanjali College of Engineering & Technology, Keesara, 501 301, Telangana, India

Abstract

$Li_2O-B_2O_3$ glasses mixed with three different modifier oxides viz., ZnO, CaO and CdO doped with Nd_2O_3 and Sm_2O_3 were prepared by melt quench method and using X-ray diffraction technique. The amorphous nature was confirmed. Differential scanning calorimetry analysis revealed reasonably good forming tendency of the glass composition. The glasses were characterized by X-ray diffraction, differential scanning calorimetry and IR spectra. FTIR spectra were used to analyze the presence of BO_3 and BO_4 functional groups in the glasses. From the optical absorption spectra, the intensities of various absorption bands of these glasses are measured and the Judd-Ofelt parameters Ω_2, Ω_4 and Ω_6 have been evaluated. The variation of Judd–Ofelt intensity parameters are discussed and correlated to the structural changes in the glass network; out of all the J-O parameters Ω_λ, the value of Ω_2, which is related to the structural changes in the vicinity of the neodymium ion indicates the highest covalent environment of Nd^{3+} ion in ZnBNd glasses and samarium ion indicates the highest covalent environment of Sm^{3+} ion in ZnBSm glasses. From this theory, various radiative properties for various emission levels of these glasses have been determined and reported. The radiative transition probabilities are evaluated from photoluminescence spectra for various luminescent transitions observed in the luminescence spectra of all the Sm^{3+} ion doped glasses suggest the highest value for $^4G_{5/2} \rightarrow {}^6H_{9/2}$ transition in ZnBSm glass.

Keywords

FTIR Spectra, Optical Absorption, Judd–Ofelt Parameters, Borate Glasses, Neodymium Ions, Samarium Ions

Contents

1. Introduction

Among the conventional glass families, borate glasses have been known to be excellent host matrices for rare earth [RE] oxides because of their good glass forming nature compared to several other conventional systems like phosphate, germanate, vanadate and tellurite glass families [1]. Optical properties of rare-earth doped glasses are extensively studied for their potential applications in the fields of lasers, fluorescent display devices, optical detectors, waveguides and fibre amplifiers [2–5]. There has been a considerable attention in the study on optical, structural and dielectric behavior of RE doped borate based glasses [6–10].

Boric acid (B_2O_3) is one of the good glass formers and can form glass alone with good transparency, high chemical durability, thermal stability and good rare-earth ion solubility [11]. Among the three modifier oxides chosen to mix in the present glass system, viz., CaO, ZnO and CdO; ZnO is expected to shorten the time taken for solidification of glasses during the quenching process and glasses containing ZnO have high chemical stability and less thermal expansion. Their wide band gap, large exciton binding energy and intrinsic emitting property make them promising candidates for the development of optoelectronic devices, solar energy concentrators, ultraviolet emitting lasers and gas sensors [12]. ZnO imparts a unique combination of optical, electrical and magnetic properties when used in glass matrices like borate. It reduces the coefficient of thermal expansion, imparts high brilliance, luster and high stability against deformation under stress [13-15] and makes the glasses nonhygroscopic and nontoxic. Both ZnO and CdO are thermally stable and appreciably covalent in character [16]. Resistance to moisture of these glasses is expected to increase by addition of alkaline-earth oxide CaO into these glass matrices [17]. The glasses mixed with Li_2O as network modifier was seen as moisture resistant, highly stable, and bubble free, is suitable for a good systematic optical analysis [18].

Neodymium ion doped glasses have attracted much attention mainly due to their advantages such as excellent mechanical and thermal stability, large solubility of rare earth ions, good capacity of glass configuration and low cost [19]. In the present work, the trivalent Neodymium ion (Nd^{3+}) has been chosen which has a $4f^2$ electronic configuration with $^4I_{9/2}$ ground state. The transition $^4I_{9/2} \rightarrow {}^2P_{1/2}$ of Nd^{3+} ion in the absorption spectra is a characteristic of the coordination of this ion. The effective coordination of this ion is found to be varying between 8 and 9 with the variations in the transition energy from 23,200-23,400 cm^{-1} [20]. The J-O theory works very well for this ion and the radiative parameters can therefore be conveniently evaluated from J-O parameters.

107

Recently, the optical properties of Nd^{3+} doped bismuth zinc borate glasses were reported by Shanmugavelu et al. [21]. Spectroscopic properties of Nd^{3+} doped borate glasses were reported by Vijaya Kumar et al. [22]. Fluorescence and radiative properties of Nd^{3+} ions doped zinc bismuth silicate glasses were reported by Pal et al. [23]. Mhareb et al. [24] reported the impact of Nd^{3+} ions on physical and optical properties of lithium magnesium borate glass. Rao et al. [25] reported the optical and structural investigation of Eu^{3+} ions in Nd^{3+} co-doped magnesium lead borosilicate glasses

Further, owing to the commercial importance of Nd^{3+} doped glass lasers, many studies on the optical properties and the structural role of Nd^{3+} ions and its interaction with the other ions in the different glass matrices have been carried out [26-28].

Further, the matrix of these glasses offers a highly suitable chemical environment for lasing ion like Sm^{3+}, since this ion can be easily incorporated homogeneously into these glasses. Samarium ion exists in Sm^{3+} and Sm^{2+} states but between these two states, Sm^{3+} is found to be more stable. This ion has a $4f^5$ electronic configuration with $^6H_{5/2}$ ground state. Earlier it was shown that the oscillator strengths of Sm^{3+} ions may be arranged in two groups, one referring to transitions up to 10,700 cm^{-1} and the second to transitions in the range of 17,600-32,800 cm^{-1} and the Judd-Ofelt parameters can be calculated separately for these two regions. Such separation was attributed to the splitting of f^N configuration being smaller than the f-d energy gap [29]. In such a case it is incorrect to use the oscillator strengths of transitions, which are about 10,000 cm^{-1} for calculations of the Ω_λ parameters by means of the Judd-Ofelt theory. The transitions $^6H_{5/2} \rightarrow {}^4F_{3/2}$, $^4F_{3/2}$ of Sm^{3+} occurring in the absorption spectrum in the near infrared region are hypersensitive [30, 31]. In the emission spectra of Sm^{3+} ion, the transitions, $^4G_{5/2} \rightarrow {}^6F_{9/2}$ and $^4G_{5/2} \rightarrow {}^6H_{9/2}$ occurring in the near infrared and visible region respectively are also identified as hypersensitive [32].

The rare earth elements are f-block elements with $4f^n5s^25p^6$ as the outer most electronic configuration when they are in the trivalent states. As mentioned earlier the volume of the rare earth ions shrinks as we go from the starting ion La^{3+} to ending ion Lu^{3+}; this shrinkage is due to the imperfect shielding of f-electrons from the nuclear charge. This shielding makes these ions to serve as active centers for laser emission and have strong bearing over the optical and electrical properties. The rare earth elements are very electropositive so their compounds are generally ionic. The arrangement of the electrons around the nuclei of the different RE elements is a determining factor of the properties of these elements. The electronic configurations of RE elements La to Lu involve the regular filling of the inner 4f shell. For the present study Nd^{3+} and Sm^{3+} ions have been chosen for the doping in $Li_2O\text{-}MO\text{-}B_2O_3$ glass matrices with a view to have an idea over the possible use of these glasses as laser hosts. For this purpose optical absorption and fluorescence properties of these glasses have been investigated. When these glasses are mixed with different network modifying ions, we may expect the structural modifications and local field variations around Nd^{3+} and Sm^{3+} ions; such changes may have strong bearing on various luminescence transitions of Nd^{3+} and Sm^{3+} ions in Lithiumborate glasses.

It is well known that the optical characterization of the glasses, i.e., the study of glass transparency, IR transmission performance and their ability to accept rare earth ions as the luminescent centers is essential for their use in the glass laser technology. During the last few years a large variety of new inorganic glasses has been developed and characterized. However most of these studies are restricted to alkali oxy borate, alkali silicate glasses. Tellurite glasses in particular are advantageous as laser hosts in view of their optical transparency over a wide range of wavelength. Transparency at shorter wavelengths of these glasses helps in getting the optimum efficiency of optical pumping of lasing ions whereas transparency in the high wavelength region facilitate to give the maximum output intensity of the laser radiation from these glasses. Certain compositions of these glasses have large rare-earth stimulated emission cross-sections and low thermo optical coefficients (compared with silicate glasses) and are the materials of choice, particularly, for high power laser applications. It is therefore felt worthwhile to investigate their optical properties after incorporating certain rare earth ions in these glasses. Studies are also extended to thermoluminescence properties of these glasses, so as to throw some light on the probable use of these glasses in radiation dosimetry.

Due to the increasing academic and technological importance of neodymium and samarium ions and the advantages of above research, the Nd^{3+} and Sm^{3+} ions doped Li_2O-MO-B_2O_3 (MO=ZnO, CaO and CdO) glasses have been prepared and investigated. Thus the objective of the present investigation is to characterize the optical absorption spectra of Nd^{3+} ions and the optical absorption and the fluorescence spectra of Sm^{3+} ions in Li_2O-MO-B_2O_3 glasses; the study is further intended to throw some light on the relationship between the structural modifications and luminescence efficiencies with the aid of IR spectral data.

2. Brief review of previous work

2.1 Brief review of previous work on glasses containing Nd^{3+} ions

Florez et al. [33] have reported compositional dependence and optical transition probabilities of Judd-Ofelt parameters of Nd^{3+} ions in flouroindate glasses. Karthikeyan et al. [34] studied the structural, optical and glass transition studies on Nd^{3+}-doped lead bismuth borate glasses. Saisudha et al. [35] have investigated the optical obsorption of

Nd^{3+}, Sm^{3+} and Dy^{3+} in bismuth borate glasses with large radiative transition probabilities.. Renuka Devi et al. [36] have analyzed the optical properties of Nd^{3+} ions in lithium borate glasses. Mehta et al. [37] studied the optical properties and spectroscopic parameters of Nd^{3+} doped phosphate and borate glasses.

Surana et al. [38] have reported the laser action in neodymium-doped zinc chloride borophosphate glasses. Jayasankar et al. [39] studied the optical properties of Nd^{3+} ions in cadmium borosulphate glasses and comparative energy level analyses of Nd^{3+} ions. Peter Tanner et al. [40] analyzed the comparative energy level parametizations for lanthanide ions in octahedral symmetry environments. Ebendorff-Heidepriem et al. [41] studied the spectroscopic properties of Nd^{3+} ions in phosphate glasses. Rao et al. [42] analyzed the luminescence properties of Nd^{3+} doped borophosphate tellurite glass. Konishi et al. [43] have reported the synthesis of ZrF_4-BaF_2,-LnF_3, glasses (Ln = La, Ce, Pr, Nd or Eu) by combined processes of sol-gel and fluorination. Ajith Kumar et al. [44] have investigated the spectroscopic parameters of Eu^{3+} and Ce^{3+} co-doped Nd^{3+} ion in phosphate glasses. De la Rosa-Cruz et al. [45] have studied the spectroscopic characterization of Nd^{3+} ions in barium flouroborophosphate glasses. Lakshman and his co-workers analysed the energy levels of Nd^{3+} ions doped binary sulphate [46] glasses by considering F^k, ξ and α parameters. Subramanyam et al. [47] studied the ternary sulphate glasses by taking into account F^k, ξ, α, β and γ parameters.

2.2   Brief review of previous work on glasses containing Sm^{3+} ions

Jayasankar and Babu [48] have reported the optical properties of Sm^{3+} ions in lithium borate and lithium fluoroborate glasses. Rodriguez et al. [49] have investigated the optical properties of Sm^{3+} doped ZnF_2-CdF_2 glasses. Reisfeld et al.. [50] have reported the absorption and emission spectral studies of Sm^{3+} doped germinate and ternary germinate glasses. Reddy et al. [51] have analyzed the absorption and emission spectral studies of Sm^{3+} and Dy^{3+} doped alkali flouoroborate glasses. Nachimuthu et al. [52] have investigated the absorption and emission spectral studies of Sm^{3+} and Dy^{3+} ions in PbO-PbF_2 glasses. Reddy et al. [53] have reported the absorption and photoluminescence spectra of some rare earth doped B_2O_3-TeO_2-BaO-R_2O (R = Li, Na, Li + Na) glasses. Sooraj Hussain et al. [54] have explained the absorption and photoluminescence spectra of Sm^{3+}:TeO_2–B_2O_3–P_2O_5–Li_2O glass. Ahrens et al. [55] have reported the determination of the Judd-Ofelt parameters of the optical transitions of Sm^{3+} in lithiumborate tungstate glasses. Biju and his co-workers [56] have investigated the energy transfer in Sm^{3+}:Eu^{3+} system in zinc sodium phosphate glasses. Jayasankar et al. [57] have reported the high-pressure fluorescence study of Sm^{3+}: lithium fluoroborate glass. Akshaya Kumar et al. [58] have analyzed the optical properties of Sm^{3+} ions doped in tellurite glass.

Schoonover et al. [59] studied the spectrum of Sm^{3+} ions in lead borate glass. Mahato et al. [60] reported the absorption, fluorescence and life-time measurements of Sm^{3+} ions in oxyfluoroborate glass.

Joshi et al. [61] have analyzed the spectra of Sm^{3+} ions doped sodium borate glasses. Sharma et al. [62] have reported the absorption spectral studies of Sm^{3+} ions doped fluorophosphate glasses. Jayasankar and Rukmini [63] have reported the optical properties of Sm^{3+} ions in zinc and alkali zinc borosulphate glasses. Souza Filho et al. [64] have analyzed the optical properties of Sm^{3+} doped lead fluoroborate glasses. Vijaya Prakash [65] has reported the absorption spectral studies of rare earth ions (Pr^{3+}, Nd^{3+}, Sm^{3+}, Dy^{3+}, Ho^{3+} and Er^{3+}) doped in NASICON type phosphate glass, $Na_4AlZnP_3O_{12}$. Takashi Kushida et al. [66] have analyzed the spectral shape of the $^5D_0 \rightarrow {}^7F_0$ line of Eu^{3+} and Sm^{2+} in glass. Annapurna et al. [67] have reported the fluorescence properties of Sm^{3+}: $ZnCl_2$-$BaCl_2$ - $LiCl$ glass. Aruna et al. [68] have reported the spectra of Sm^{3+} and Dy^{3+}: B_2O_3 - P_2O_5 - $R_2 SO_4$ glasses. Jianrong Qiu et al. [69] have analyzed the photostimulated luminescence in borate glasses doped with Eu^{2+} and Sm^{3+} ions.

3. Aim of the present work

Li_2O-MO-B_2O_3 glasses have high thermal expansion coefficient and have high mechanical strength and are transparent to both UV and visible regions. These glasses can easily be prepared and were considered as the good materials for applications in battery sealing and for enamel paints. Though some studies on certain physical properties of lithium borate glasses are available in the literature, the detailed systematic investigations on electrical, optical properties and thermoluminescence properties of Li_2O-MO-B_2O_3: Nd^{3+} and Sm^{3+} glasses are not available. Detailed studies on optical properties viz., optical absorption and fluorescence of rare earth doped Li_2O-MO-B_2O_3 glasses will throw some light on the possible use of these glasses as laser hosts. X-ray irradiation of especially rare-earth ions doped Li_2O-MO-B_2O_3 glasses may produce different types of color centers in these glasses. The information about the color centers in these glasses may be obtained by studying optical absorption and thermoluminescence of the X-ray irradiated glasses. These studies may help for considering the applications of these glasses in radiation dosimeters etc.

4. Details of the present investigation

The details of the present investigation dealing with the optical properties of the Lithiumborate glasses are:

i) Differential scanning calorimetric studies and the evaluation of glass transition temperature T_g.

ii) The optical absorption spectra of the glasses were recorded at room temperature in the wavelength range 300-900 nm.

iii) By using xenon arc lamp, the intense line λ_{exc} = 400 nm was identified and the same was used to record photoluminescence spectra in the UV and visible region at room temperature.

iv) Infrared spectra of all these glasses in the region 400 to 4000 cm^{-1} and to study the effect of rare-earth ions on the position and intensity of various vibrational bands.

v) Thermoluminescence in the temperature range 30 °C - 300 °C after the glasses are X-ray irradiated and evaluation of various trap depth parameters.

5.   Experimental methods

5.1   Preparation of glasses

Undoped and following Eu^{3+} ion doped glasses are prepared by using standard melting and quenching techniques and used for the present study. [70-72].

Borate Pure Series:	ZnB:	30 Li_2O - 10 ZnO - 60 B_2O_3
	CaB:	30 Li_2O - 10 CaO - 60 B_2O_3
	CdB:	30 Li_2O - 10 CdO - 60 B_2O_3
Borate Nd^{3+} Series:	ZnBNd:	30 Li_2O - 10 ZnO - 59 B_2O_3: 1 Nd_2O_3
	CaBNd:	30 Li_2O - 10 CaO - 59 B_2O_3: 1 Nd_2O_3
	CdBNd:	30 Li_2O - 10 CdO - 59 B_2O_3: 1 Nd_2O_3
Borate Sm^{3+} Series:	ZnBSm:	30 Li_2O - 10 ZnO - 59 B_2O_3: 1 Sm_2O_3
	CaBSm:	30 Li_2O - 10 CaO - 59 B_2O_3: 1 Sm_2O_3
	CdBSm:	30 Li_2O - 10 CdO - 59 B_2O_3: 1 Sm_2O_3

Appropriate amounts of raw materials ZnO, $CaCO_3$, CdO, H_3BO_3, Li_2CO_3, Nd_2O_3 and Sm_2O_3 (all in mole %) were thoroughly mixed and grounded in an agate mortar and melted in a platinum crucible. High purity (99.9%) chemicals were used in this work. All these chemicals with the said compositions were heated in a PID temperature controlled furnace at 450°C for 2 hour for the decorbonization from $CaCO_3$ and Li_2CO_3. Then the temperature was raised to the range of 1000-1050°C and kept at this temperature for an hour till a bubble free liquid was formed. For the homogeneous mixing of the constituents the crucibles were shaken frequently. The resultant melt was poured on a rectangular brass mold held at room temperature. The samples were subsequently annealed at glass transition temperature in another furnace to remove mechanical stress and were then polished. Bubble free and optically transparent glasses were selected for optical studies.

5.2 X–ray diffraction

The amorphous state of the prepared glasses was checked by X-ray diffraction spectra recorded on "PHILIPS EXPERT" Diffractometer having copper target with nickel filter operated at 40 kV, 30 mA. The absence of peaks in the spectra indicates the amorphous nature of the materials. Glassy materials do not have a long-range atomic order, i.e., atoms are arranged randomly. Therefore, a diffraction pattern containing sharp peaks is not expected as observed in crystalline materials.

5.3 Differential scanning calorimetry

The glass transition temperatures T_g and crystallization temperature T_c of these glasses were determined (to an accuracy of $\pm$ 1°C) by differential scanning calorimetry (DSC) traces, recorded using universal V23C TA differential scanning calorimeter with a programmed heating rate of 15°C per minute in the temperature range 30-750°C.

5.4 Optical absorption spectra

The optical absorption spectra of the glasses were recorded at room temperature in the wavelength range of 300-900 nm for neodymium ion doped glasses and 300-500 nm for samarium ion doped glasses using Shimadzu-3100 UV-VIS-NIR Spectrophotometer. By using a xenon arc lamp, the intense line λ_{exc} = 400 nm was identified and the same was used to record the photo-luminescence spectrum of samarium ion doped glasses.

5.5 Photoluminescence spectra

The photo-luminescence spectra of the glasses were recorded with a Hitachi-F 3010 Fluorescence Spectrophotometer in the wavelength range of 500-700 nm up to a resolution of 0.1 nm.

5.6 Infrared spectroscopy

Infrared transmission spectra for these glasses were recorded using a Perkin Elmer Spectrometer in the wavenumber range of 400-4000 cm^{-1} by KBr pellet method.

6. Judd-Ofelt theory

It is well known that there is a shielding of the 4f electrons of the rare-earth ions and this shielding allow these ions to serve as active centres in solid state laser hosts. These ions exhibit sharp absorption and luminescence transitions as surrounding ligand atoms weakly perturb them. The spectral intensities for the observed bands of these glasses that are often expressed in terms of oscillator strength of forced electronic dipole transitions have been analyzed with the help of the Judd-Ofelt theory [73, 74] using:

$$f = \frac{8\pi^2 mv}{3h(2J+1)} \frac{(n_d^2+2)^2 v}{9n_d} \sum_{\lambda=2,4,6} \Omega_\lambda \left(\psi_J \| U^\lambda \| \psi'_{J'}\right)^2 \tag{1}$$

where (2J+1) is the multiplicity of the lower states, m is the mass of the electron and 'v' is the peak absorption in cm^{-1}. Experimental values of oscillator strengths were evaluated from the expression:

$$f_{exp} = 2.302 \left(\frac{mc^2}{N_A \pi e^2}\right) \int \varepsilon(v) dv \tag{2}$$

where N_A is the Avogadro's number and ε(v) is the molar absorption coefficient.

Using J-O parameters Ω_λ, the radiative properties of fluorescent transitions from ^5D_o level for the present glasses are determined. The spontaneous emission probability A, is calculated using the expression:

$$A(\psi_J, \psi'_{J'}) = \frac{64\pi^4 e^2 v^3}{3h(2J'+1)} \left[\frac{n_d(n_d^2+2)^2}{9} S_{ed} + n_d^3 S_{md}\right] \tag{3}$$

where

$$S_{ed} = \sum_{\lambda=2,4,6} \Omega_\lambda \left(\psi_J \| U^\lambda \| \psi'_{J'}\right)^2 \tag{4}$$

and

$$S_{md} = \frac{e^2 h^2}{16\pi^2 m^2 c^2} \left(\psi_J \| L + 2S \| \psi'_{J'}\right)^2 \tag{5}$$

In the above equation, (2J+1) is the multiplicity of the upper state and 'v' is the wavenumber of the fluorescence peak. It may be noted here that the contribution from the magnetic dipoles to the transition probability A for the specific transitions evaluated is found to be either zero or almost negligible.

Then the total emission probability A_T involving all the intermediate terms is calculated using:

$$A_T(\psi_J) = \sum_{\psi'_{J'}} A(\psi_J, \psi'_{J'}) \tag{6}$$

The radiative lifetime (τ_r) of a state is calculated using the relationship:

$$\tau_R = 1/A_T(\psi_J) \tag{7}$$

The fluorescent branching ratio is obtained from the equation:

$$\beta_r(\psi_J, \psi'_{J'}) = \frac{A(\psi_J, \psi'_{J'})}{A_T(\psi_J)} \tag{8}$$

Finally the stimulated emission cross sections of the measured fluorescent levels are evaluated using:

$$\sigma_P = \frac{A(\psi_J, \psi'_{J'})\lambda^4}{8\pi c n^2_a \Delta\lambda} \tag{9}$$

where λ is the peak position of the emission line and $\Delta\lambda$ is the effective band width of the emission transitions.

7.   Results and discussion

7.1   Characterization

The existence of glass transition temperature T_g and crystallization temperature T_c in differential scanning calorimetry (DSC) study curves and absence of peaks in X-ray diffraction pattern indicate that the glasses prepared were of high quality glasses.

Fig.1A shows the differential scanning calorimetry traces of Nd^{3+} ions doped glasses and Fig 1B shows the differential scanning calorimetry traces of Sm^{3+} ions doped glasses. All Nd^{3+} ions doped glasses exhibit an endothermic change between 533°C and 548.3°C, and all Sm^{3+} ions doped glasses exhibit an endothermic change between 533°C and 555°C; which is attributed to the glass transition temperature T_g. At still higher temperature, T_c an exothermic peak due to the crystal growth followed by another endothermic effect at temperature T_m due to the re-melting of the glass are also observed. The appearance of single peak due to the glass transition temperature in DSC pattern of all the glasses indicates the high homogeneity of the glasses prepared. From the measured values of T_g, T_c and T_m, the parameters, T_g/T_m, $(T_c-T_g)/T_g$, $(T_c-T_g)/T_m$ and glass forming ability parameter known as Hruby's parameter $K_{gl} = (T_c-T_g)/(T_m-T_c)$, are evaluated and presented in Table 1. The highest values of these parameters are obtained for ZnO-modifier glass doped with Sm^{3+} ions (ZnBSm) indicating it's relatively high glass forming ability among all the glasses.

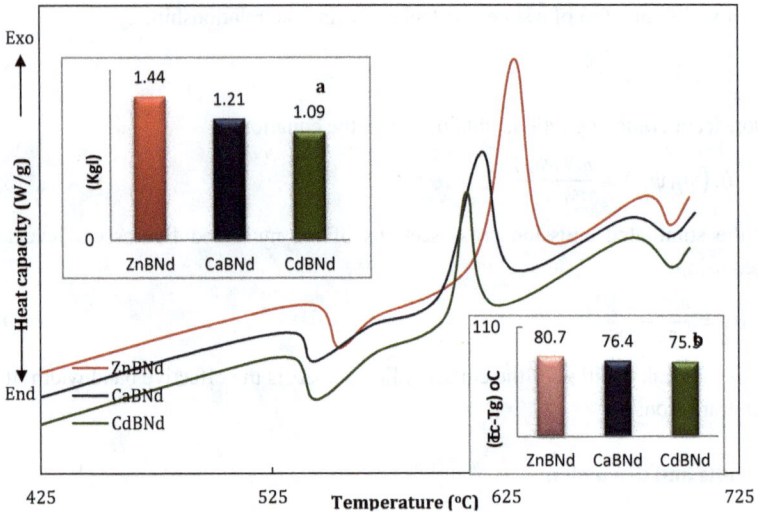

Fig.1A: DSC patterns of Nd^{3+} doped Li$_2$O-MO-B$_2$O$_3$ glasses. Insets represent the variation of a) Hruby's parameter and b) (T$_c$-T$_g$) values for different modifiers.

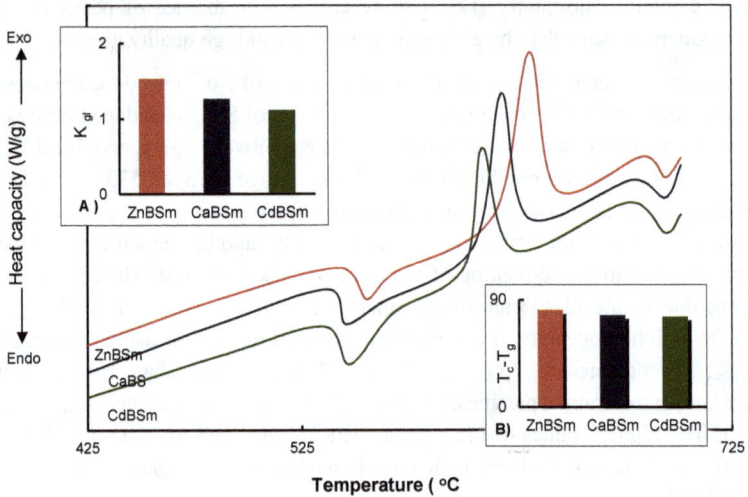

Fig.1B: DSC patterns of Sm^{3+} doped Li$_2$O-MO-B$_2$O$_3$ glasses. Insets represent the variation of a) Hruby's parameter and b) (T$_c$-T$_g$) values for different modifiers.

Table 1 Data on differential scanning calorimetric studies of Li_2O-MO-B_2O_3: Nd_2O_3 and Sm_2O_3 glasses.

Glass	T_g (°C)	T_c (°C)	T_m (°C)	T_g/T_m	$(T_c$-$T_g)$	$(T_c$-$T_g)/T_m$	K_{gl}
ZnBNd	548.3	629	685.2	0.800	80.7	0.118	1.436
CaBNd	538.6	615	678	0.794	76.4	0.113	1.213
CdBNd	533.0	608.5	678	0.786	75.5	0.111	1.086
ZnBSm	549.8	631	684	0.804	81.2	0.119	1.532
CaBSm	541.0	618	679	0.797	77.0	0.113	1.262
CdBSm	533.3	609	677	0.788	75.7	0.112	1.113

7.2 Optical absorption spectra

The optical absorption spectra recorded at room temperature for Nd^{3+} doped glasses have exhibited the following absorption bands and Fig. 2A shows the absorption spectra of Nd^{3+} ion in Li_2O-ZnO-B_2O_3 glasses.

$$^4I_{9/2} \rightarrow {}^4F_{3/2}, {}^4F_{5/2}, {}^4F_{7/2}, {}^4F_{9/2}, {}^4G_{5/2}, {}^4G_{7/2}, {}^4G_{9/2}, {}^4G_{11/2}, {}^2P_{1/2} \text{ and } {}^4D_{5/2}$$

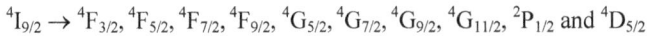

The optical absorption spectrum recorded at room temperature in the visible region for Sm^{3+} doped glasses has exhibited the following eight prominent absorption bands. Fig. 2B shows the optical absorption spectra of Sm^{3+} doped Li_2O-MO-B_2O_3 glasses.

$$^6H_{5/2} \rightarrow {}^4I_{11/2}, {}^4I_{13/2}, {}^4G_{9/2}, {}^6P_{5/2}, {}^4F_{7/2}, {}^6P_{7/2}, {}^4D_{5/2}, {}^4K_{17/2}$$

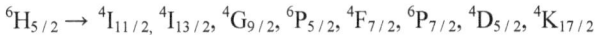

From these absorption spectral profiles, it is observed that a particular transition $^6H_{5/2} \rightarrow {}^4F_{7/2}$ is more intense than any other transition. The trends of the observed intensities of transitions found in all three glasses are similar. All above transitions are based on the assignments of lanthanide spectra reported by Carnall et al. [75]. For these observed bands, the Judd-Ofelt theory is applied to characterize the spectral intensities of the absorption bands in these glasses. The best fit Judd-Ofelt intensity parameters Ω_λ and the oscillator strengths of Nd^{3+} and Sm^{3+} doped Li_2O-MO-B_2O_3 glasses are presented in Table 2A and Table 2B respectively. From these absorption spectral profiles, it is observed that a particular transition $^4I_{9/2} \rightarrow {}^4G_{5/2}$ is more intense than any other transition in Nd^{3+} doped glasses and the transition $^4F_{3/2} \rightarrow {}^4I_{11/2}$ is more intense than any other transition in Sm^{3+} doped glasses. This is obviously because of better validity of the selection rules: $\Delta J < 2$, $\Delta L < 2$ and $\Delta S = 0$, for this transition. In addition, the magnitude of $\|U^\lambda\|^2$ of this level is also considered as an important value for the hypersensitivity nature of this level [76]. There is a reasonable agreement between the experimental and calculated values of oscillator strengths (Tables 2A & 2B).

Table 2A Measured f_{exp} (10^{-6}) and calculated f_{cal} (10^{-6}) oscillator strengths of Nd^{3+} doped Li_2O-MO-B_2O_3 glasses

Transition	ZnBNd		CaBNd		CdBNd	
from $^4I_{9/2}$	f_{exp}	f_{cal}	f_{exp}	f_{cal}	f_{exp}	f_{cal}
$^4F_{3/2}$	2.21	1.41	2.98	1.90	2.78	1.42
$^4F_{5/2}$	2.44	3.25	2.65	3.50	2.56	3.01
$^4F_{7/2}$	2.52	1.25	2.74	1.05	2.61	1.06
$^4F_{9/2}$	0.42	0.31	0.46	0.29	0.45	0.27
$^4G_{5/2}$	1.49	9.73	1.77	10.58	1.65	8.94
$^4G_{7/2}$	1.85	2.17	2.17	2.66	2.37	2.11
$^4G_{9/2}$	1.49	0.87	1.75	1.06	1.03	0.84
$^4G_{11/2}$	0.09	0.12	0.10	0.13	0.09	0.11
$^2P_{1/2}$	0.26	0.95	2.36	2.01	1.02	2.56
r.m.s deviation	±1.0540		± 1.1168		± 0.9445	

Table 2B. Measured f_{exp} (10^{-6}) and calculated f_{cal} (10^{-6}) oscillator strengths of Sm^{3+} doped Li_2O-MO-B_2O_3 glasses

Transition	ZnBSm		CaBSm		CdBSm	
from $^6H_{5/2}$	f_{exp}	f_{cal}	f_{exp}	f_{cal}	f_{exp}	f_{cal}
$^4I_{11/2}$	0.20	0.29	0.16	0.23	0.19	0.29
$^4I_{13/2}$	0.69	0.72	0.54	0.59	0.72	0.77
$^4G_{9/2}$	0.08	0.11	0.07	0.09	0.09	0.12
$^6P_{5/2}$	0.17	0.78	0.14	0.66	0.18	0.96
$^4F_{7/2}$	0.15	0.06	0.13	0.05	0.16	0.07
$^6P_{7/2}$	2.49	2.38	2.01	1.92	2.62	2.47
$^4D_{5/2}$	0.33	0.59	0.20	0.50	0.18	0.72
r.m.s deviation	±0.0834		± 0.0505		± 0.1225	

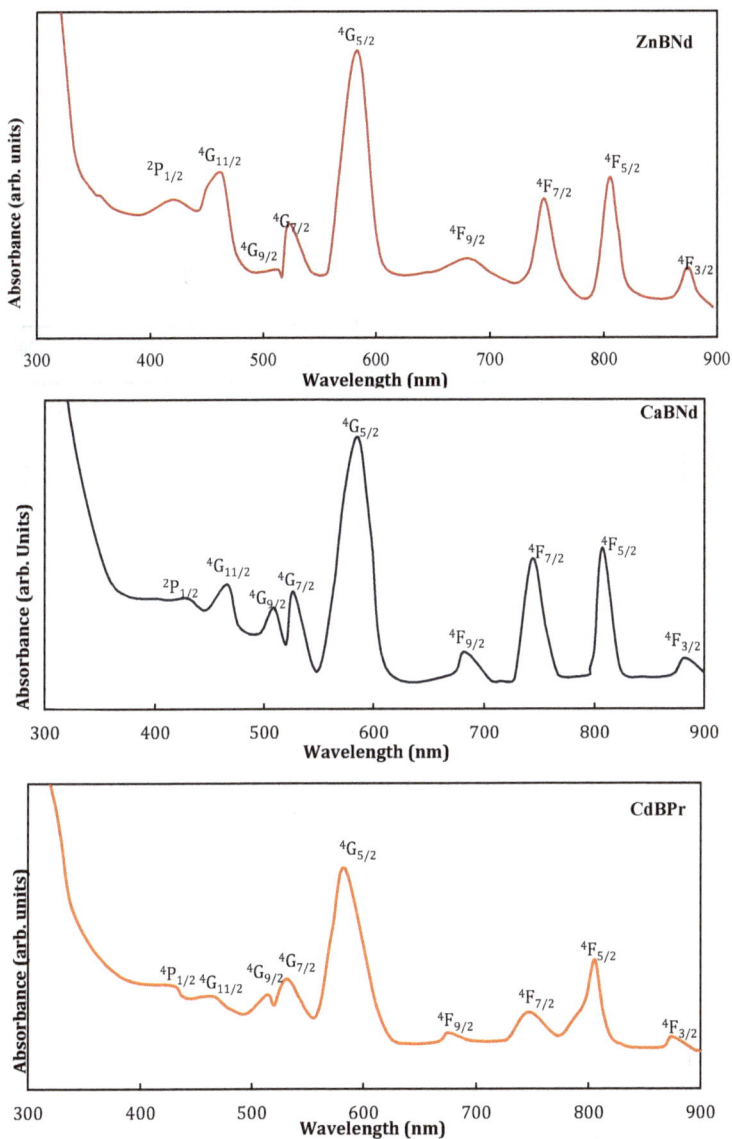

Fig. 2A: Optical absorption spectra of Nd^{3+} doped Li_2O-ZnO-B_2O_3 glasses. All transitions are from the ground state $^4I_{9/2}$.

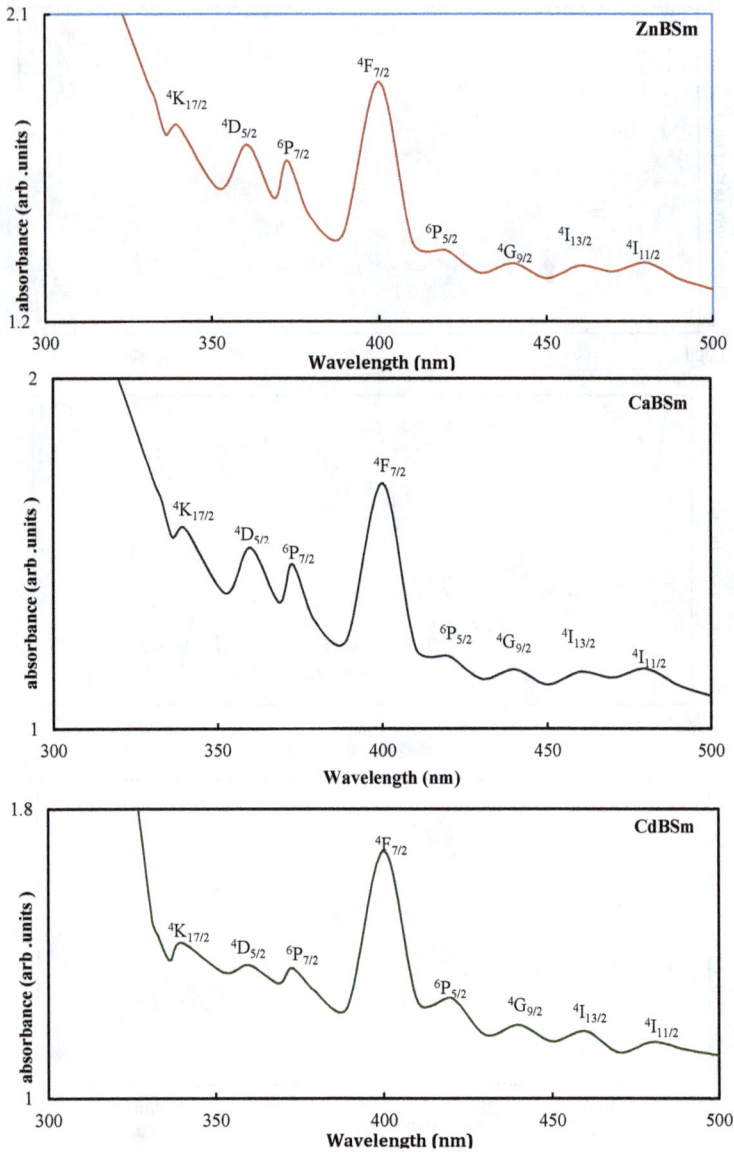

Fig.2B: *Optical absorption spectra of Li_2O-MO-B_2O_3 glasses doped with Sm^{3+} ions recorded at room temperature. All the transitions are from the ground state $^6H_{5/2}$.*

Similar trends are observed in the intensities of transitions found in all three glasses of each series. The r.m.s deviations of oscillator strengths of experimental and calculated values are presented in Tables 2A & 2B. The observed relatively small values of these deviations confirm the validity and applicability of Judd-Ofelt theory for the present glasses.

Table 3 Judd-Ofelt parameters $\Omega_\lambda x10^{-20}$ (cm^{-2}), spectroscopic quality factor and the bonding parameter (δ') of Nd_2O_3 and Sm_2O_3 ions doped Lithiumborate glasses.

Glass	Ω_2	Ω_4	Ω_6	SQF (Ω_4/Ω_6)	δ'
ZnBNd	4.47	4.82	4.24	1.14	0.4809
CaBNd	4.16	7.05	3.54	1.99	0.4357
CdBNd	3.87	5.01	3.60	1.39	0.4136
ZnBSm	5.77	19.34	17.97	1.08	0.1200
CaBSm	5.46	16.62	14.67	1.13	0.1063
CdBSm	5.33	16.12	12.56	1.28	0.1047

The Judd-Ofelt parameters Ω_2 , Ω_4 and Ω_6 are computed by the least square fitting analysis of the experimental oscillator strengths using matrix elements and spectroscopic quality factor (SQF) of all the glasses presented in Table 3. The values of Ω_λ show the following order for all the Nd^{3+} doped glasses: $\Omega_4 > \Omega_2 > \Omega_6$ and for all the Sm^{3+} doped glasses: $\Omega_4 > \Omega_6 > \Omega_2$. The comparison of J-O parameters shows the highest value of Ω_2 for the ZnBNd Nd^{3+} doped glasses and the highest value of Ω_2 for ZnBSm Sm^{3+} doped glasses. The bonding parameter, δ' [77, 78] is also calculated for all the glasses. The computation shows the highest δ' value for ZnBNd Nd^{3+} doped glasses and ZnBSm Sm^{3+} doped glasses.

The parameter Ω_2 is related to the covalency and structural changes in the vicinity of the Nd^{3+} ion in Nd^{3+} doped glasses and Sm^{3+} ion in Sm^{3+} doped glasses (short-range effect) and Ω_4 and Ω_6 are related to the long-range effects. The comparison of Ω_2 parameter for the Sm^{3+} doped glasses (Table 4) shows the highest value for Zn-O modifier glasses, indicating the highest covalent character of this glass. The larger modifier ion (Cd^{2+} ionic radius, 1.03 Å) give rise to a large average distance between the BO$_4$ chains which results in the average RE-O distance to increase, therefore producing a weaker field around the rare earth ion leading to a low value of Ω_2 when compared with that of Ca-O modifier glasses (Ca^{2+} ionic radius, 0.99 Å) and Zn-O modifier glasses (Zn^{2+} ionic radius, 0.74 Å).

A similar conclusion can also be drawn from the values of Ω_6, which is related to the rigidity of the host [85]. Further support for this argument can also be cited from the value of the bonding parameter δ', the value of δ' for these glasses follows the order CdBNd<CaBNd<ZnBNd indicating the high covalent environment for ZnBNd glass in Nd^{3+} ions doped glasses and follows the order CdBSm<CaBSm<ZnBSm indicating the high covalent environment for ZnBSm glass in Sm^{3+} ions doped glasses.

Table 4 Comparision of Judd-Ofelt intensity parameter values $\Omega_\lambda (10^{-20}$ cm^2) of Nd^{3+} ions doped glass systems.

Glass type	Ref.	Ω_2	Ω_4	Ω_6	SQF (Ω_4/Ω_6)	Trend
ZnBNd	Present work	4.47	4.82	4.24	1.14	$\Omega_4 > \Omega_2 > \Omega_6$
CaBNd	Present work	4.16	7.05	3.54	1.99	$\Omega_4 > \Omega_2 > \Omega_6$
CdBNd	Present work	3.87	5.01	3.60	1.39	$\Omega_4 > \Omega_2 > \Omega_6$
CaLiBO	[79]	7.85	6.75	7.61	0.89	$\Omega_2 > \Omega_6 > \Omega_4$
SrLiBO	[79]	6.52	5.00	6.16	0.81	$\Omega_2 > \Omega_6 > \Omega_4$
25 Bi_2O_3 - 25PbO - 49 B_2O_3	[34]	1.577	4.483	3.333	1.35	$\Omega_4 > \Omega_6 > \Omega_2$
97 [30 Bi_2O_3 : 70B_2O_3]	[34]	4.646	2.898	5.854	0.50	$\Omega_6 > \Omega_2 > \Omega_4$
33.3ZnO–66.6TeO$_2$	[80]	4.24	0.88	7.05	0.12	$\Omega_6 > \Omega_2 > \Omega_4$
BZBNd10	[21]	2.67	3.31	3.98	0.83	$\Omega_6 > \Omega_4 > \Omega_2$
BINLAB5	[22]	1.30	4.09	5.23	0.78	$\Omega_6 > \Omega_4 > \Omega_2$
GeO$_2$–PbO–Bi$_2$O$_3$	[81]	2.95	5.01	3.93	1.27	$\Omega_4 > \Omega_6 > \Omega_2$
LGBNd	[82]	4.84	5.97	4.59	1.30	$\Omega_6 > \Omega_2 > \Omega_4$
NGBNd	[82]	5.75	3.44	3.73	0.92	$\Omega_2 > \Omega_6 > \Omega_4$
TZN10	[83]	3.80	4.94	4.54	1.09	$\Omega_4 > \Omega_6 > \Omega_2$
TLF	[84]	5.61	4.17	5.44	0.77	$\Omega_2 > \Omega_6 > \Omega_4$

7.3 Photoluminescence spectra

In the emission spectrum, the focus in general is mainly on $^4F_{3/2} \rightarrow {}^4I_{11/2}$, which is identified as the lasing transition. However in the present study the emission spectrum for this ion in the present host glass could not be carried out due to the lack of a proper laser excitation source.

From the optical absorption spectral profiles, it is observed that a particular transition $^6H_{5/2} \rightarrow {}^4F_{7/2}$ is more intense than any other transition in Sm^{3+} ions doped glasses. The trends of the observed intensities of transitions found in Sm^{3+} ions doped glasses are similar. The room temperature fluorescence spectra of all the Sm^{3+} ions doped glasses recorded in the wavelength region 500-700 nm with the excited wavelength 400 nm have exhibited different emission bands (Fig. 3) identified due to the following transitions

$$^4G_{5/2} \rightarrow {}^6H_{5/2}, {}^6H_{7/2}, {}^6H_{9/2}.$$

The values of the transition probability $A(\psi_J, \psi_{J'})$, the total transition probability $A_T(\psi_J)$ and the fluorescence branching ratio β_r evaluated using the equations mentioned in section 6 for the transitions originated from $^4G_{5/2}$ levels are presented in Table 5.

Table 5. Radiative properties of Sm^{3+} doped Li_2O-MO-B_2O_3 glasses

Glass System	Emission Transition	A (s^{-1}) x (10^4)	A_T (s^{-1}) x (10^4)	τ_R (μs)	β_R (%)	Emission cross section σ_Px10^{20} (cm^2)
ZnBSm	$^4G_{5/2} \rightarrow {}^6H_{9/2}$	1.28	1.95	51.3	65.83	8.39
	$^4G_{5/2} \rightarrow {}^6H_{7/2}$	0.67	1.95	51.3	34.17	1.92
	$^4G_{5/2} \rightarrow {}^6H_{5/2}$	0.50	1.70	58.8	29.42	0.79
CaBSm	$^4G_{5/2} \rightarrow {}^6H_{9/2}$	1.20	1.83	54.71	65.81	9.29
	$^4G_{5/2} \rightarrow {}^6H_{7/2}$	0.61	1.83	54.71	33.64	3.76
	$^4G_{5/2} \rightarrow {}^6H_{5/2}$	0.50	1.70	58.82	29.42	0.88
CdBSm	$^4G_{5/2} \rightarrow {}^6H_{9/2}$	1.11	1.74	57.53	64.04	7.93
	$^4G_{5/2} \rightarrow {}^6H_{7/2}$	0.62	1.74	57.53	35.39	3.49
	$^4G_{5/2} \rightarrow {}^6H_{5/2}$	0.50	1.70	58.82	29.42	0.84

The radiative properties of Sm^{3+} ions (or any of Ln^{3+} ions) depend on the number of factors such as network former and modifier of the glass. The parameter β_r (i.e. the branching ratio) of the luminescence transitions characterizes the lasing power of the potential laser transitions. It is well established that an emission level with β_r value nearly equal to 50% becomes a potential laser emission [86]. Referring to the data on emission transitions in the present glass systems, the transition $^4G_{5/2} \rightarrow {}^6H_{9/2}$ has the highest value of β_r for all the Sm^{3+} ions doped glasses glasses among various transitions; this transition may therefore be considered as possible laser transition. However, the comparison of β_r values of this transition for these glasses shows that the value of β_r is largest for ZnBSm glass indicating its better suitability for lasing action among the Sm^{3+} ions doped glasses (Table 4).

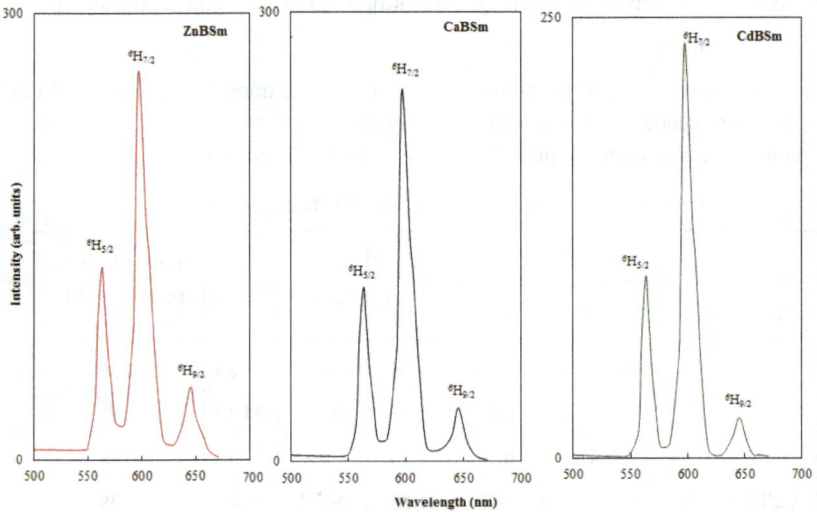

Fig.3: Photoluminescence spectra of Sm^{3+} doped Li_2O-MO-B_2O_3 glasses (I_{exc} = 400 nm) recorded at room temperature. All the transitions are from the upper state $^4G_{5/2}$.

The predicted life time (τ_R) of $^4G_{5/2}$ for CdBSm glass is observed to be larger than those of CaBSm and ZnBSm glasses indicating, the smaller interaction of Sm^{3+} ion in CdB glass network. This may be due to the higher ionic radius of the Cd compared to the Zn and Ca [87].

7.4 Infrared spectroscopy

Infrared transmission (IR) spectra of the pure as well as Nd^{3+} and Sm^{3+} doped Li_2O-MO-B_2O_3 glasses are shown in Fig. 4. The infrared transmission spectra of Nd^{3+} doped glasses exhibit three groups of bands: (i) in the region 1320-1360 cm^{-1}, (ii) in the region 960-1020 cm^{-1} and (iii) a band at about 710 cm^{-1}. The infrared transmission spectra of pure glasses and Sm^{3+} ion doped glasses exhibit three groups of bands: (i) in the region 1200-1600 cm^{-1}, (ii) in the region 800-1200 cm^{-1} and (iii) a band at about 710 cm^{-1}.

It is well known that the effect of introduction of alkali oxides into B_2O_3 glass is the conversion of sp^2 planar BO_3 units into more stable sp^3 tetrahedral BO_4 units and may also create non-bridging oxygens. Each BO_4 unit is linked to two such other units and one oxygen from each unit with a rare earth ion and the structure leads to the formation of long tetrahedron chains. The second group of bands is attributed to such BO_4 units whereas the first group of bands is identified as due to the stretching relaxation of the B-O bond of the trigonal BO_3 units and the band at 710 cm^{-1} is due to the bending vibrations of B-O-B linkages in the borate network [88-91]. The weak band observed around 456 cm^{-1} is an indicative of the presence of ZnO_4 units in the zinc borate glass network [92,93].

Table 6 Peak positions (cm^{-1}) of IR spectra of Nd^{3+} and Sm^{3+} doped Li_2O-MO-B_2O_3 glasses.

Glass	Band due to B-O bond stretching in BO_3 units	Band due to B-O bond stretching in BO_4 units	Band due to B-O-B linkage in borate network
ZnBNd	1360	963	710
CaBNd	1336	992	710
CdBNd	1323	1019	710
ZnBSm	1353	975	710
CaBSm	1327	1002	710
CdBSm	1316	1027	710

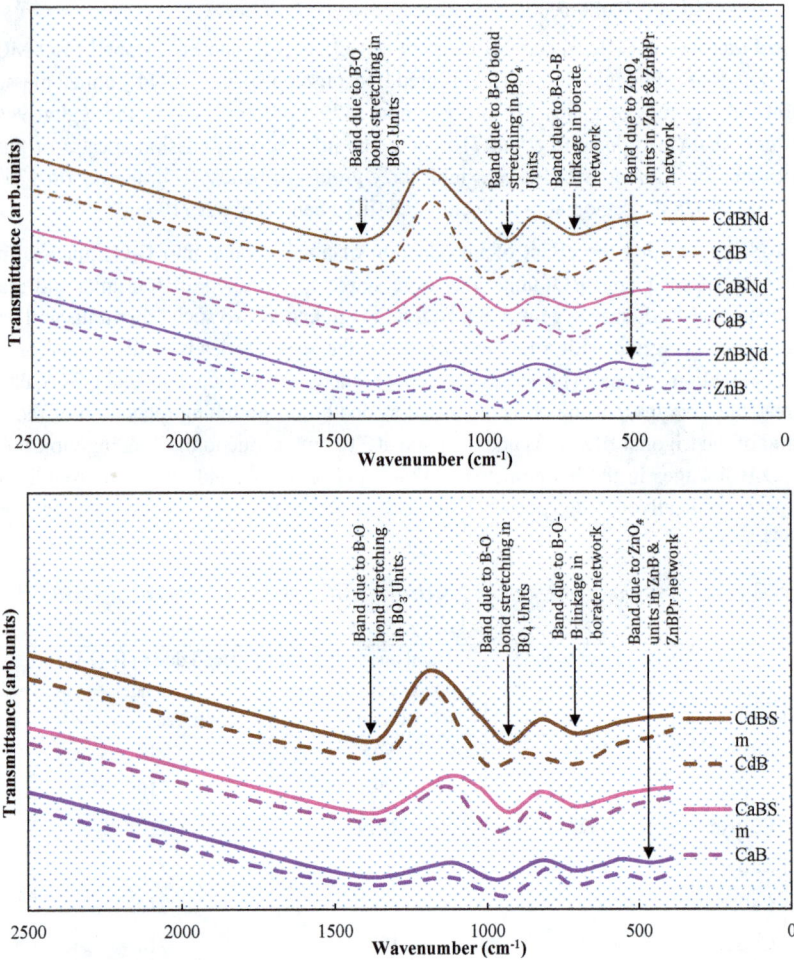

Fig. 4: Infrared transmission (IR) spectra of the pure as well as Nd^{3+} and Sm^{3+} doped Li_2O-MO-B_2O_3 glasses

When the glasses are doped with Nd_2O_3 and Sm_2O_3, the intensity of the second group of bands (band due to the trigonal BO_4 units) is found to increase at the expense of first group of bands (bands due to tetrahedral BO_3 units) with the increase of atomic number of rare earth ions with the shifting of meta-centres of first and second group of bands, respectively towards slightly lower and higher wave number for all the glasses. No significant change in position and intensity of the other bands is observed in the spectra of all the series of the glasses by introducing the Nd^{3+} ion or Sm^{3+} ion. The summary of the data on the positions of various bands in the IR spectra of Nd^{3+} ion and Sm^{3+} ion doped Li_2O-MO-B_2O_3 glasses are presented in Table 6.

8. Conclusions

The influence of modifier oxide on spectroscopic characteristics of Li_2O-MO-B_2O_3 glasses doped with Nd_2O_3 and Sm_2O_3 doped Li_2O-MO-B_2O_3 glasses has been studied.

i) Differential scanning calorimetric studies indicate a high glass forming ability for ZnBNd glass in Nd^{3+} ion doped glasses and high glass forming ability for ZnBSm glass in Sm^{3+} ion doped glasses. ZnBSm glass has a high glass forming ability in all the glass samples. ii) The IR spectral studies indicate relatively less disorder in the ZnBSm glass network. iii) The variation of the Judd–Ofelt intensity parameters are discussed and correlated to the structural changes in the glass network; out of all the J-O parameters Ω_λ, the value of Ω_2, which is related to the structural changes in the vicinity of the neodymium ion indicates the highest covalent environment of Nd^{3+} ion in ZnBNd glasses and samarium ion indicates the highest covalent environment of Sm^{3+} ion in ZnBSm glasses. (iv) The radiative transition probabilities evaluated for various luminescent transitions observed in the luminescence spectra of all the Sm^{3+} ion doped glasses suggest the highest value for $^4G_{5/2} \rightarrow ^6H_{9/2}$ transition in ZnBSm glass.

References

[1]   Yamane M, Asahara Y. Glasses for Photonics, Cambridge University Press, 2000. http://dx.doi.org/10.1017/CBO9780511541308

[2]   Balakrishna A, Rajesh, D, Ratnakaram Y.C. Structural and optical properties of Nd^{3+} in lithium fluoro-borate glass with relevant modifier oxides. Optical Materials, 2013, 35 (12): 2670. http://dx.doi.org/10.1016/j.optmat.2013.08.004

[3]   Ying Guan, Zhihao Wei, Yanlin Huang, Ramzi Maalej, Hyo Jin Seo. 1.55 μm emission and upconversion luminescence of Er^{3+}- doped strontium borate glasses. Ceramics International, 2013, 39: 7023. http://dx.doi.org/10.1016/j.ceramint.2013.02.040

[4]   Anjaiah J, Laxmikanth C, Veeraiah N. Spectroscopic properties and luminescence behaviour of europium doped lithium borate glasses. Physica B, 2014, 454: 148. http://dx.doi.org/10.1016/j.physb.2014.07.070

[5]   Susumu Nakayama, Takamitsu Watanabe, Taro Asahi, Hajime Kiyono, Yan Lin Aung, Masatomi Sakamoto. Influence of rare earth additives and boron component on electrical conductivity of sodium rare earth borate glasses. Ceramics International, 2010, 36: 2323. http://dx.doi.org/10.1016/j.ceramint.2010.07.026

[6]   Ramteke DD, Annapurna K, Deshpande VK, Gedam RS. Effect of Nd^{3+} on spectroscopic properties of lithium borate glasses. Journal of Rare Earths, 2014, 32(12), 1148. http://dx.doi.org/10.1016/S1002-0721(14)60196-4

[7]   Bhushana Reddy M, Sailaja S, Nageswara Raju C, Sudhakar Reddy B. Optical absorption and NIR emission spectral studies of Er^{3+} and Nd^{3+}: zinc lithium bismuth borate glasses. Journal of Optics, 2014, 43: 101. http://dx.doi.org/10.1007/s12596-014-0189-6

[8]   Ramteke DD, Gedam RS. Impedance spectroscopic characterization of Sm_2O_3 containing lithium borate glasses. Spectrochimica Acta Part A: Molecular and Biomolecular Spectroscopy, 2014, 133: 19. http://dx.doi.org/10.1016/j.saa.2014.05.005

[9]   Kashif I, Abd El-Maboud A, Ratep. A. Effect of Nd_2O_3 addition on structure and characterization of lead bismuth borate glass. Results in Physics, 2014, 4: 1. http://dx.doi.org/10.1016/j.rinp.2013.11.002

[10]   Swapna K, Mahamuda Sk, Srinivasa Rao A, Jayasimhadri M, Sasikala T, Rama Moorthy L. Visible fluorescence characteristics of Dy^{3+} doped zinc alumino bismuth borate glasses for optoelectronic devices. Ceramics International, 2013, 39: 8459. http://dx.doi.org/10.1016/j.ceramint.2013.04.028

[11]   Yan Zhang, Chunhua Lu, Liyan Sun, Zhongzi Xu, Yaru Ni. Influence of Sm_2O_3 on the crystallization and luminescence properties of boroaluminosilicate glasses, Materials Research Bulletin, 2009, 44: 179. http://dx.doi.org/10.1016/j.materresbull.2008.03.004

[12]   Annapurna K, Das M, Kundu M, Dwivedhi R.N, Buddhudu S. Spectral properties of Eu^{3+}: ZnO–B_2O_3–SiO_2 glasses, Journal of Molecular Structure. 2005, 741: 53. http://dx.doi.org/10.1016/j.molstruc.2005.01.062

[13]   Pinckney LR. Transparent glass-ceramics based on ZnO Crystals, Physics and Chemistry of Glasses. European Journal of Glass Science and Technology Part B, 2006, 47: 127.

[14]   Miguel A, Morea R, Gonzalo J, Arriandiaga MA, Fernandez J, Balda R. Near-infrared emission and upconversion in Er^{3+}-doped TeO_2–ZnO–ZnF_2 glasses,

Journal of Luminescence, 2013, 140: 38.
http://dx.doi.org/10.1016/j.jlumin.2013.02.059

[15]   Ticha H, Kincl M, Tichy L. Some structural and optical properties of
       ,$Bi_2O_{3,x}$,$ZnO_{,60-x}$,$B_2O_{3,40}$ glasses, Materials Chemistry and *Physics,* 2013, 138:
       633. http://dx.doi.org/10.1016/j.matchemphys.2012.12.032

[16]   Lee JD. Concise Inorganic Chemistry, Oxford: Blackwell Science, 1996.

[17]   Brahmaiah A, Bala Murali Krishna S, Kondaiah M, Bala Narendra Prasad T,
       Krishna Rao D. Influence of chromium ions on dielectric and spectroscopic
       properties of Na_2O-PbO-B_2O_3glass system, IOP Conference Series: Materials
       Science and Engineering, 2009, 2: 012028. http://dx.doi.org/10.1088/1757-
       899X/2/1/012028

[18]   Rao TVR, Reddy RR, Nazeer Ahammed Y, Parandamaiah M., et. al.
       Luminescence properties of Nd^{3+}, TeO_2-B_2O_3–P_2O_5–Li_2O glass. Infrared Physics
       & Technology 2000, 41: 247. http://dx.doi.org/10.1016/S1350-4495(99)00060-2

[19]   Serqueira EO, Dantas NO, Monte AFG, Bell MJV. Judd Ofelt calculation of
       quantum efficiencies and branching ratios of Nd^{3+} doped glasses, Journal of Non-
       Crystalline Solids, 2006, 352: 3628.
       http://dx.doi.org/10.1016/j.jnoncrysol.2006.03.093

[20]   Srinivasa Rao L, Srinivasa Reddy M, Ramana Reddy M.V, Veeraiah N.
       Spectroscopic features of Pr^{3+}, Nd^{3+}, Sm^{3+} and Er^{3+} ions in Li_2O–MO ,Nb_2O_5,
       MoO_3 and WO_3,–B_2O_3 glass systems, Physica B: Condensed Matter, 2008, 403:
       2542. http://dx.doi.org/10.1016/j.physb.2008.01.043

[21]   Shanmugavelu B, Venkatramu V, Ravi Kanth Kumar VV. Optical properties of
       Nd^{3+} doped bismuth zinc borate glasses, Spectrochimica Acta Part A: Molecular
       and Biomolecular Spectroscopy, 2014, 122: 422.
       http://dx.doi.org/10.1016/j.saa.2013.11.051

[22]   Vijaya Kumar K, Suresh Kumar A. Spectroscopic properties of Nd^{3+} doped borate
       glasses, Optical Materials, 2012, 35(1): 12.
       http://dx.doi.org/10.1016/j.optmat.2012.06.005

[23]   Pal I, Agarwal A, Sanghi S, Aggarwal MP, Bhardwaj S. Fluorescence and
       radiative properties of Nd^{3+} ions doped zinc bismuth silicate glasses, Journal of
       Alloys and Compounds, 2014, 587: 332.
       http://dx.doi.org/10.1016/j.jallcom.2013.10.191

[24] Mhareb MHA, Hashim S, Ghoshal SK, Alajerami YSM, Saleh MA, Dawaud RS, Razak NAB, Azizan SAB. Impact of Nd^{3+} ions on physical and optical properties of Lithium Magnesium Borate glass, Optical Materials, 2014, 37: 391. http://dx.doi.org/10.1016/j.optmat.2014.06.033

[25] Rao TGVM, Rupesh Kumar A, Neeraja K, Veeraiah N, Rami Reddy M. Optical and structural investigation of Eu^{3+} ions in Nd^{3+} co-doped magnesium lead borosilicate glasses, Journal of Alloys and Compounds, 2013, 557: 209. http://dx.doi.org/10.1016/j.jallcom.2012.12.162

[26] Campbell JH, Suratwala TI. Nd-doped phosphate glasses for high-energy/high-peak-power lasers, Journal of Non-Crystalline Solids, 2000, 263–264: 318. http://dx.doi.org/10.1016/S0022-3093(99)00645-6

[27] Bettinell, M, Speghini A, Brik MG. Spectroscopic studies of emission and absorption properties of $38PbO-62SiO_2:Nd^{3+}$ glass, Optical Materials, 2010, 32: 1592. http://dx.doi.org/10.1016/j.optmat.2010.05.027

[28] Gandhi Y, Kityk IV, Brik M.G, Raghava Rao P, Veeraiah N. Influence of tungsten on the emission features of Nd^{3+}, Sm^{3+} and Eu^{3+} ions in $ZnF_2-WO_3-TeO_2$ glasses, Journal of Alloys and Compounds, 2010, 508: 278. http://dx.doi.org/10.1016/j.jallcom.2010.08.137

[29] Reisfeld, R., 1975, Structure and Bonding 22 "Springer- verlag, New York", 123

[30] Henrie, D.E., Fellows, R.L., Choppin, G.R., 1976, Coord. Chem. Rev. 18, p. 199. http://dx.doi.org/10.1016/S0010-8545(00)82044-5

[31] Peacock, R.D., Structure and Bonding 1975, "Springer-Verlag, New York", 22, p.83

[32] Horrocks, W.D., Albin, M., 1984, "Prog. Inorg. Chem" (An Inter science Publication, New York,) 31, pp.1

[33] Florez A., Martinez J.F., Florez M., J.Non-Crystlline Solids 284 (2001) 261. http://dx.doi.org/10.1016/S0022-3093(01)00412-4

[34] Karthikeyan B., Mohan S., Physica B 334 (2003) 298. http://dx.doi.org/10.1016/S0921-4526(03)00080-2

[35] Saisudha M.B., Ramakrishna J., Optical Materials 18 (2002) 403. http://dx.doi.org/10.1016/S0925-3467(01)00181-1

[36] Renuka Devi A., Jayasankar C.K., Materials Chemistry and Physics 42 (1995) 106. http://dx.doi.org/10.1016/0254-0584(95)01564-7

[37]   Mehta V., Aka G., Dawar A.L., Mansingh A., Optical Materials 12 (1999) 53. http://dx.doi.org/10.1016/S0925-3467(98)00074-3

[38]   Surana S.S.L., Sharma1 Y.K., Tandon S.P.,Materials Science and Engg. B 83 (2001) 204. http://dx.doi.org/10.1016/S0921-5107(01)00517-7

[39]   Jayasankar C.K., Ravi Kanth Kumar V.V., Physica B 226 (1996) 313. http://dx.doi.org/10.1016/0921-4526(96)00288-8

[40]   Peter A. Tanner., Kumar V.V.R.K., Jayasankar C.K., J. Alloys and compounds 225 (1995) 85. http://dx.doi.org/10.1016/0925-8388(94)07014-8

[41]   Ebendorff-Heidepriem H., Seeber W., Ehrt D., Journal of Non-Crystalline Solids 183 (1995) 191. http://dx.doi.org/10.1016/0022-3093(94)00560-5

[42]   Rao T.V.R., Reddy R.R., Nazeer Ahammed Y., Parandamaiah M., Infrared Physics & Technology 41 (2000) 247. http://dx.doi.org/10.1016/S1350-4495(99)00060-2

[43]   Konishi A., Kanno R., Kawamoto Y., J. All. and Comp. 232 (1996) 53. http://dx.doi.org/10.1016/0925-8388(95)02020-9

[44]   Ajith Kumar G., Biju P.R., Venugopal C., Journal of Non-Crystalline Solids 221 (1997) 47. http://dx.doi.org/10.1016/S0022-3093(97)00266-4

[45]   De la Rosa-Cruz E., Kumar G.A., Diaz-Torres L.A., Optical Materials 18 (2001) 321. http://dx.doi.org/10.1016/S0925-3467(01)00171-9

[46]   Lakshman S.V.J. and Ratnakaram Y.C., Phys. Chem. Glasses, 29 (1988) 26.

[47]   Subramanyam Y., Moorthy L.R. and Lakshman S.V.J., J. Phys. Chem. Solids 51 (1990) 231. http://dx.doi.org/10.1016/0022-3697(90)90106-P

[48]   Jayasankar C.K., Babu P., J. Alloy. and Comp. 307 (2000) 82. http://dx.doi.org/10.1016/S0925-8388(00)00888-4

[49]   Rodriguez V.D., Martin I.R., Alcala R., Cases R., J. Lumin. 54 (1992) 231. http://dx.doi.org/10.1016/0022-2313(92)90070-P

[50]   Reisfeld R., Bornstein A., Boehm L., J. Solid State Chem. 14 (1975) 14. http://dx.doi.org/10.1016/0022-4596(75)90356-4

[51]   Reddy R.R., Nazeer Ahammed Y., Abdul Azeem P., Rama Gopal K., Rao T.V.R., Buddhudu S., Sooraj Hussain N., Journal of Quantitative Spectroscopy & Radiative Transfer 77 (2003) 149. http://dx.doi.org/10.1016/S0022-4073(02)00084-5

[52]   Nachimuthu P., Jagannathan R., Nirmal Kumar V., Narayana Rao D., Journal of Non-Crystalline Solids 217 (1997) 215. http://dx.doi.org/10.1016/S0022-3093(97)00151-8

[53]   Reddy C.V., Ahammed Y.N., Reddy R.R. and Rao T.V.R., J. Phys. Chem. Solids 159 (1998) 337. http://dx.doi.org/10.1016/S0022-3697(97)00210-2

[54]   Sooraj Hussain N., Aruna V., Buddhudu S., Materials Research Bulletin, 35 (2000) 703.

[55]   Ahrens H., Wollenhaupt M., Frobel P., Jun Lin, Barner K., Sun G.S., Braunstein R., Journal of Luminescence 82 (1999) 177. http://dx.doi.org/10.1016/S0022-2313(99)00051-4

[56]   Biju P.R., Jose G., Thomas V., Nampoori V.P.N., Unnikrishnan N.V., Optical Materials 24 (2004) 671. http://dx.doi.org/10.1016/S0925-3467(03)00183-6

[57]   Jayasankar C.K., Babu P., Troster Th., Holzapfel W.B., J. Luminescence 91 (2000) 33. http://dx.doi.org/10.1016/S0022-2313(00)00206-4

[58]   Akshaya Kumar, Rai D.K., Rai S.B., Spectrochimica Acta Part A 59 (2003) 917. http://dx.doi.org/10.1016/S1386-1425(02)00282-2

[59]   Schoonover J.R., Lee Y.L., Su S.N., Lin S.H., Eyring L., Appl. Spectrosc. 28 (1984) 154. http://dx.doi.org/10.1366/0003702844554099

[60]   Mahato K.K., Rai D.K., Rai S.B., Solid State Commun. 108 (1998) 671. http://dx.doi.org/10.1016/S0038-1098(98)00442-6

[61]   Joshi J.C., Joshi J., Belwel R., Joshi B.C., Pandey N.C., J.Phys. Chem. Solids, 39 (1978) 581. http://dx.doi.org/10.1016/0022-3697(78)90040-9

[62]   Sharma B., Rai S.B., Rai D.K., Buddhudu S., Indian J.Eng. Mater. Sci. 2 (1995) 297.

[63]   Jayasankar C.K., Rukmini E., Optical Materials 8 (1997) 193. http://dx.doi.org/10.1016/S0925-3467(97)00021-9

[64]   Souza Filho A.G., Mendes Filho J., Melo F.E.A., Custo dio M.C.C., Lebullenger R., Hernandes A.C., Journal of Physics and Chemistry of Solids 61 (2000) 1535. http://dx.doi.org/10.1016/S0022-3697(00)00032-9

[65]   Vijaya Prakash G., Materials Letters 46 (2000) 15. http://dx.doi.org/10.1016/S0167-577X(00)00135-X

[66]   Takashi Kushida, Atusi Kurita, Masanori Tanaka., Journal of Luminescence 102 (2003) 301. http://dx.doi.org/10.1016/S0022-2313(02)00519-7

[67] Annapurna K., Dwevedi R.N., Kundu P., Buddhudu S., Materials Research Bulletin 38 (2003) 429. http://dx.doi.org/10.1016/S0025-5408(02)01068-1

[68] Aruna V., Sooraj Hussain N., Buddhudu S., Materials Research Bulletin, 33 (1998) 149.

[69] Jianrong Qiu, Shimizuguwa Y., Sugimoto N., Hirao K., Journal of Non-Crystalline Solids 222 (1997) 290. http://dx.doi.org/10.1016/S0022-3093(97)90126-5

[70] A. Paul, Chemistry of glasses, Chapman & Hall, London (1982). http://dx.doi.org/10.1007/978-94-009-5918-7

[71] S.R. Elliot, Physics of amorphous materials, (Longman, London 1990).

[72] J. F. Shackl Ford, Introduction to Materials Science for Engineers, (Macmillan, New York, 1985).

[73] B. R. Judd, Phys. Rev 127 (1962) 750. http://dx.doi.org/10.1103/PhysRev.127.750

[74] G.S. Ofelt, J. Chem. Phys. 37 (1962) 511. http://dx.doi.org/10.1063/1.1701366

[75] Carnall WT, Feilds PR, Rajnak K. Electronic energy levels in the trivalent lanthanide aquo ions. I. Pr^{3+}, Nd^{3+}, Pm^{3+}, Sm^{3+}, Dy^{3+}, Ho^{3+}, Er^{3+}, and Tm^{3+}, The Journal of Chemical Physics, 1968, 49: 4424. http://dx.doi.org/10.1063/1.1669893

[76] Lucas J. Rare earths in fluoride glasses, Journal of the Less-Common Metals, 1985, 112: 27. http://dx.doi.org/10.1016/0022-5088(85)90005-0

[77] Jorgenson CK. Orbitals Atoms and Molecules, Academic Press, London, 1962.

[78] Sinha SP. Complexes of the Rare Earths, Pergamon, Oxford, 1966. http://dx.doi.org/10.1016/b978-0-08-011616-7.50005-x

[79] Renuka Devi A, Jayasankar CK. Optical properties of Nd^{3+} ions in lithium borate glasses, Materials chemistry and physics, 1995, 42: 106. http://dx.doi.org/10.1016/0254-0584(95)01564-7

[80] Kanoun A, Alaya S, Maaref H. Spectroscopic properties of Pr^{3+} and Nd^{3+} ions in zinc tellurite glasses, Physica Status Solidi B. 1990, 162: 523. http://dx.doi.org/10.1002/pssb.2221620224

[81] Balda R, Fernandez J, Sanz M, de Pablos A, Fdez-Navarro JM, Mugnier J. Infrared-to-visible upconversion in Nd^{3+}-doped chalcohalide glasses , Physical Review B , 2000, 61: 144101.

[82] Sunil Kumar S, Jayakrishna Khatei Kasthurirengan S, Koteswara Rao K.S.R, Ramesh, K.P. Optical absorption and photoluminescence studies of Nd^{3+} doped

alkali boro germanate glasses, Journal of Non-Crystalline Solids, 2011, 357: 842. http://dx.doi.org/10.1016/j.jnoncrysol.2010.11.007

[83] Upendra Kumar K, Prathyusha VA, Babu P, Jayasankar CK, Joshi AS, Speghini A, Bettinelli M. Fluorescence properties of Nd^{3+}-doped tellurite glasses, Spectrochimica Acta Part A, 2007, 67: 702. http://dx.doi.org/10.1016/j.saa.2006.08.027

[84] Ravi Kanth Kumar VV, Bhatnagar A, Jagannathan R. Structural and optical studies of Pr^{3+}, Nd^{3+}, Er^{3+} and Eu^{3+} ions in tellurite based oxyfluoride, TeO_2-LiF, glass, Journal of Physics D: Applied Physics, 2001, 34: 1563. http://dx.doi.org/10.1088/0022-3727/34/11/301

[85] Vermelho MVD, Neto ASG, Amorim HT, Cassanjes FC, Ribeirod SJL, Messaddeqd Y. Temperature investigation of infrared-to-visible frequency upconversion in erbium-doped tellurite glasses excited at 1540 nm, Journal of Luminescence, 2003, 102: 755. http://dx.doi.org/10.1016/S0022-2313(02)00637-3

[86] Adam, J. L., Sibley, W. A., 1985, "J. Non-Cryst. Solids" 76, pp.267-279. http://dx.doi.org/10.1016/0022-3093(85)90004-3

[87] Hirayama, C., Camp, F.E., Melamid, N.T., Steinbruegge, K.B., 1971 "J. Non-Cryst. Solids" 6, pp.342-356. http://dx.doi.org/10.1016/0022-3093(71)90025-1

[88] Tandon RP, Hotchandani S. Electrical conductivity of semiconducting tungsten oxide glasses. Physica Status Solidi (a) 2001, 185: 453. http://dx.doi.org/10.1002/1521-396X(200106)185:2<453::AID-PSSA453>3.0.CO;2-L

[89] Qiu HH, Sakata H, Hirayama T. Electrical conduction of glasses in the system Fe_2O_3-Sb_2O_3-TeO_2, Journal of Ceramic Society of Japan, 1995, 103: 32. http://dx.doi.org/10.2109/jcersj.103.32

[90] Khalifa FA, Azooz A. Infrared absorption spectra of gamma irradiated glasses of the system Li_2O-B_2O_3-Al_2O_3. Indian Journal of Pure & Applied Physics, 1998, 36: 314.

[91] Ahmed AA, Eltohamy MR. Infrared study of glasses in the system B_2O_3-PbO-CuO in relation to their structure. Indian Journal of Pure & Applied Physics, 1998, 36: 335.

[92] Subbalakshmi P, Veeraiah N. Influence of WO_3 on dielectric properties of zinc phosphate glasses. Indian Journal of Engineering & Materials Sciences, 2001, 8: 275.

[93]   Karthikeyan B, Mohan, Baesso M.L. Spectroscopic and glass transition studies on Nd^{3+}-doped sodium zincborate glasses. Physica B , 2003, 337: 249. http://dx.doi.org/10.1016/S0921-4526(03)00411-3

CHAPTER 6

Relationship between the structural modifications and luminescence efficiencies of ZnF_2-MO-TeO_2 glasses doped with Ho^{3+} and Er^{3+} ions

C. Laxmikanth[1], J. Anjaiah[1,2]

[1] Department of Physics, The University of Dodoma, Tanzania, East Africa

[2] Department of Physics, Geethanjali College of Engineering & Technology, Keesara, 501 301, Telangana, India

Abstract

ZnF_2-MO-TeO_2 (where MO stands for ZnO, CdO and PbO) glasses doped with Ho_2O_3 and Er_2O_3 ions were prepared by melt quench method. The characterization of these glasses was done by characterized by XRD and IR spectra. The infrared spectral studies indicate relatively less disorder in ZTHo glass network in Ho^{3+} doped glasses and ZTE glass network in Er^{3+} doped glasses. The room temperature optical absorption and fluorescence spectra of these glasses have been studied and presented. The Judd-Ofelt parameters Ω_2, Ω_4 and Ω_6 of these glasses have been evaluated from the measured intensities of various absorption bands and compared the same with those of other reported glass systems. The Judd-Ofelt theory was successfully applied in evaluating the various radiative parameters for different emission levels of these glasses and the same were reported. The highest values of the branching ratios are found to be for the emission transitions $^5S_2 \rightarrow {}^5I_8$ in ZTHo glass and $^4F_{5/2} \rightarrow {}^4I_{15/2}$ in CTE, respectively, among holmium (where the transitions originating from $^5G_5$, $^5S_2$ and $^5F_3$ levels) and erbium doped glasses (where the transitions originating from $^2G_{9/2}$ and $^4F_{5/2}$ levels).

Keywords

Optical Absorption, Fluorescence, Infrared Spectra, Holmium Ions, Erbium Ions, Tellurite Glasses

Contents

1.   Introduction

Tellurite glasses were considered as the best materials for optical components such as IR domes, optical filters, modulators, memories and laser windows in view of their high transparency in the far infrared region and for their high density and refractive index. Further, these glasses are considered as great materials for hosting lasing rare-earth ions since these materials provide a low phonon energy environment to minimize non-radiative losses. A number of recent investigations on mechanical, electrical and optical properties of various tellurite glasses mixed with different modifiers are available in the literature [1-8]. A considerable number of these studies on various physical properties have also been reported in the recent past from our laboratory on some TeO_2 based glass systems [9-14]. The addition of ZnF_2 into TeO_2 glass matrices lowers the viscosity and is expected to decrease the liquidus temperature to a substantial extent and further it acts as an effective mineralizer [15]. Among the three modifier oxides, viz., ZnO, CdO and PbO, added to the glass matrix, ZnO is expected to shorten the time taken for the solidification of glasses during the quenching process. Both ZnO and CdO are thermally stable, sublime and appreciably covalent in character [16]. The addition of PbO into the glass matrix produces low rates of crystallization, since PbO has the ability to form stable glasses due to its dual role; one as glass former when Pb-O bonding is covalent and the other as modifier when bonding is ionic [17]. In view of these qualities, all the three modifier oxides are interesting oxides and make the glasses more stable against devitrification and resistant to moisture.

Among various rare earth ions, Ho^{3+} ion exhibits several electronic transitions in the visible and infrared regions. As a consequence there are many laser transitions in its emission spectrum; this ion exhibits eye safe potential laser even at room temperature with a low threshold action [18-20] that have attractive applications in atmospheric communication systems. A quite good number of recent studies on the lasing action of Ho^{3+} ion in various glass matrices are available in the literature [21,22]. Further when the glasses are mixed with different network modifying ions we may expect the structural modifications and local field variations around Ho^{3+} ion; such changes may have strong bearing on various luminescence transitions of Ho^{3+} ions in ZnF_2-MO-TeO_2 glasses (where MO stands for PbO, ZnO and CdO).

Another rare earth ion, Er^{3+} ion has a number of strong absorption bands where the pumping sources are easily available. As a consequence the emission spectrum exhibits many fluorescent lines, blue, green and red in the visible region. Excellent luminescence efficiencies have been reported in a variety of inorganic glass systems containing Er^{3+} ions [23-27]. The laser oscillation of this ion has also been utilized as a fiber amplifier of silica at 1.55μm [28,29]. This ion exhibits eye safe potential laser even at room temperature with a low threshold action [30,31] that have attractive applications in atmospheric communication systems. Further when these glasses are mixed with different network modifying ions, we may expect the structural modifications and local field variations around Er^{3+} ion; such changes may have strong bearing on various luminescence transitions of Er^{3+} ions in the glass matrix.

Thus the objective of the present investigation is to characterize the optical absorption and the fluorescence spectra of Ho^{3+} ions and Er^{3+} ions in ZnF_2-TeO_2 glasses mixed with three different modifier oxides viz., ZnO, CdO and PbO; the study is further intended to throw some light on the relationship between the structural modifications and luminescence efficiencies with the aid of IR spectral data.

2. Experimental methods

For the present investigation the following compositions have been chosen:

ZnF_2-MO-TeO_2: Ho_2O_3 glass series:

ZTHo glass: $40ZnF_2$–$10ZnO$-$49TeO_2$: $1Ho_2O_3$,

CTHo glass: $40ZnF_2$–$10CdO$-$49TeO_2$: $1Ho_2O_3$,

PTHo glass: $40ZnF_2$–$10PbO$-$49TeO_2$: $1Ho_2O_3$.

ZnF_2-MO-TeO_2: Er_2O_3 glass series:

ZTE glass: $40ZnF_2$–$10ZnO$-$49TeO_2$: $1Er_2O_3$,

CTE glass: $40ZnF_2$–$10CdO$-$49TeO_2$: $1Er_2O_3$,

PTE glass: $40ZnF_2$–$10PbO$-$49TeO_2$: $1Er_2O_3$.

Appropriate amounts (all in wt.%) of analar grade reagents of TeO_2 (99.99% pure, Aldrich), PbO, ZnO, CdO, ZnF_2, Ho_2O_3 and Er_2O_3 were thoroughly mixed for respective glass series in an agate mortar and melted in a platinum crucible between 630-680°C in a PID temperature controlled furnace for about ½ hr until a bubble free liquid was formed. The resultant melt was then cast in a brass mold and subsequently annealed at 200°C. The weight losses were found to be less than 0.5%. The amorphous state of the glasses was checked by X-ray diffraction studies.

The densities ρ of the glasses were determined to an accuracy of 0.001 g/cm^3 by standard principle of Archimedes' using O-xylene (99.99% pure) as the buoyant liquid. The samples were then ground and optically polished. The final dimensions of the samples used for the present measurements were about 1 cm x 1 cm x 0.2 cm. The optical absorption spectra of the glasses were recorded at room temperature in the wavelength range 370-2100 nm using Shimadzu-UV-VIS-NIR Spectrophotometer Model 3100. By using xenon arc lamp, the intense line $\lambda_{exc.}$ = 450 nm was identified for Ho^{3+} glasses and the same was used to record the photo-luminescence spectrum. In case of Er^{3+} glasses the intense line $\lambda_{exc.}$ = 380 nm was identified and the same was used to record the photo-luminescence spectrum of all glasses. The photo-luminescence spectra of the glasses were recorded on Hitachi - F 3010 Fluorescence Spectrophotometer in the wavelength range 460-700 nm for Ho^{3+} doped glasses and in the wavelength range 380-840 nm for Er^{3+} doped glasses up to a resolution of 0.1 nm. Infrared transmission spectra for these glasses were recorded using a Perkin Elmer Spectrometer in the wavenumber range 400-4000 cm^{-1} by KBr pellet method.

3. Results

From the measured values of the density ρ and calculated average molecular weight $\overline{M}$, various physical parameters such as holmium ion concentration N_i, mean holmium ion separation R_i that are useful for understanding optical properties of Ho^{3+} doped glasses, are evaluated and presented in Table 1A and in a similar way, various physical properties of Er^{3+} doped glasses are evaluated and presented in Table 1B.

Table 1A Various physical properties of ZnF_2-MO-TeO_2: Ho_2O_3 glasses.

Property	ZTHo	CTHo	PTHo
Refractive index, n_d	1.569	1.571	1.570
Density, ρ (g/cm^3)	5.591	5.618	5.646
Average molecular weight, $\bar{M}$	121.4	128.5	122
Holmium ion concentration, N_i (10^{19}/cm^3)	20.4	20.5	8.8
Inter-ionic distance of holmium ions, R_i (Å)	16.98	16.96	22.46

Table 1B Various physical properties of ZnF_2-MO-TeO_2:Er_2O_3 glasses

Property	ZTE	CTE	PTE
Refractive index, n_d	1.570	1.575	1.578
Density, ρ (g/cm^3)	5.602	5.620	5.632
Average molecular weight, $\bar{M}$	122.01	129	135.97
Holmium ion concentration, N_i (10^{19}/cm^3)	8.82	8.84	8.88
Inter-ionic distance of holmium ions, R_i (Å)	22.46	22.44	22.41

Fig. 1A shows the optical absorption spectra of ZnF_2-MO-TeO_2 glass containing Ho_2O_3, recorded at room temperature in the wavelength region 400-2100 nm. The spectra exhibited eight clearly resolved absorption bands all from the ground state $^5I_8$; these levels are identified due to the following electronic transitions from the holmium ions [32]:

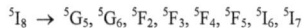

$$^5I_8 \rightarrow {}^5G_5, {}^5G_6, {}^5F_2, {}^5F_3, {}^5F_4, {}^5F_5, {}^5I_6, {}^5I_7$$

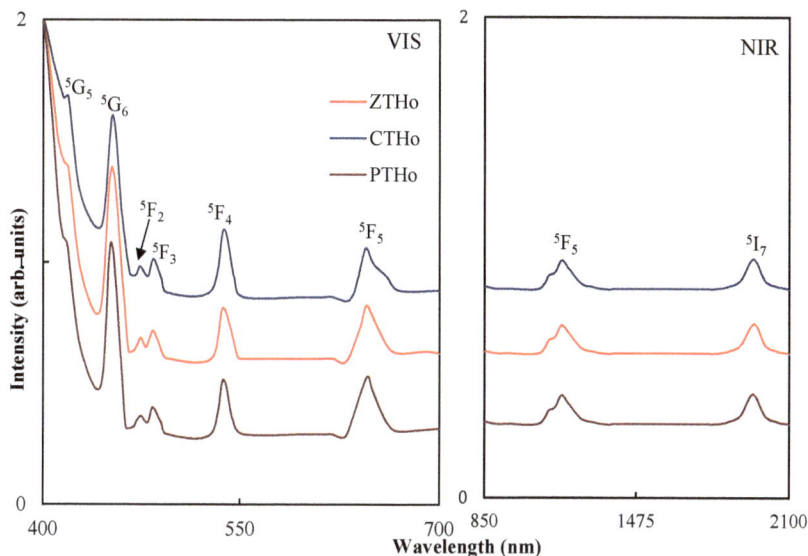

Figure 1A. Optical absorption spectra of Ho^{3+} doped ZnF$_2$-MO-TeO$_2$ glasses recorded at room temperature (all the transitions are from the ground state $^5I_8$)

Fig. 1B shows the optical absorption spectra of ZnF$_2$-MO-TeO$_2$ glasses containing Er$_2$O$_3$, recorded at room temperature in the wavelength region 370-2100 nm. The spectra exhibited eight clearly resolved absorption bands out of fourteen levels predicted, all from the ground state $^4I_{15/2}$; these levels are identified due to the following electronic transitions [32]:

$$^4I_{15/2} \rightarrow {}^4G_{11/2}, {}^4F_{7/2}, {}^2H_{11/2}, {}^4S_{3/2}, {}^4F_{9/2}, {}^4I_{9/2}, {}^4I_{11/2}, {}^4I_{13/2}.$$

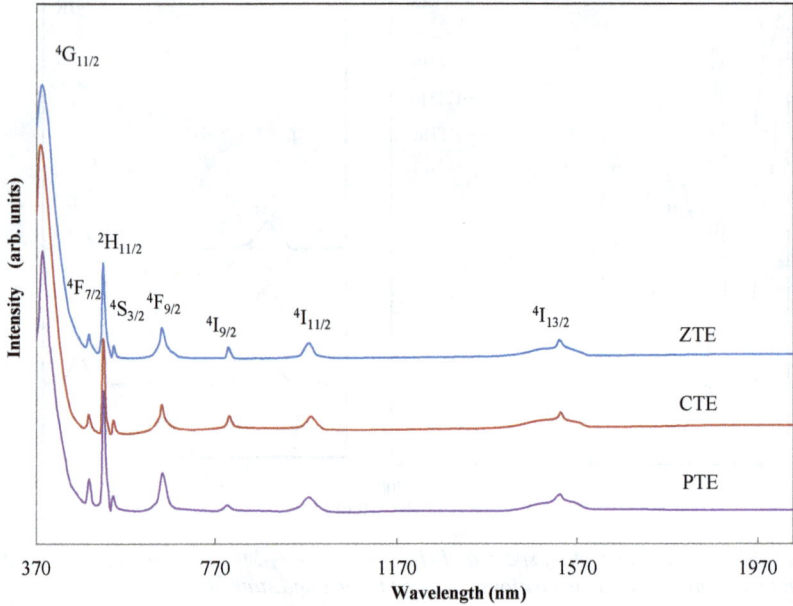

Figure 1B. Optical absorption spectrum of Er^{3+} doped ZnF_2-MO-TeO_2 glass (ground state $^4I_{15/2}$)

Further, the trends of the observed intensities of transitions found in all Ho^{3+} doped glasses are similar. The room temperature fluorescence spectrum of ZnF_2-MO-TeO_2 glass doped with Ho^{3+} ions recorded in the wavelength region 460-700 nm with the excited wavelength 450 nm have exhibited different emission bands (Fig. 2A) identified due to the following transitions:

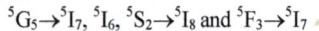

$$^5G_5 \rightarrow {}^5I_7, \; {}^5I_6, \; {}^5S_2 \rightarrow {}^5I_8 \; \text{and} \; {}^5F_3 \rightarrow {}^5I_7$$

The room temperature fluorescence spectra of all the Er^{3+} doped glasses recorded in the wavelength region 380-840 nm with the excited wavelength 380 nm have exhibited different emission bands (Fig. 2B) identified due to the following transitions [32]:

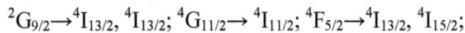

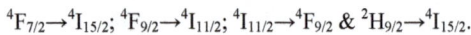

$$^2G_{9/2} \rightarrow {}^4I_{13/2}, \; {}^4I_{13/2}; \; {}^4G_{11/2} \rightarrow {}^4I_{11/2}; \; {}^4F_{5/2} \rightarrow {}^4I_{13/2}, \; {}^4I_{15/2};$$

$$^4F_{7/2} \rightarrow {}^4I_{15/2}; \; {}^4F_{9/2} \rightarrow {}^4I_{11/2}; \; {}^4I_{11/2} \rightarrow {}^4F_{9/2} \; \& \; {}^2H_{9/2} \rightarrow {}^4I_{15/2}.$$

142

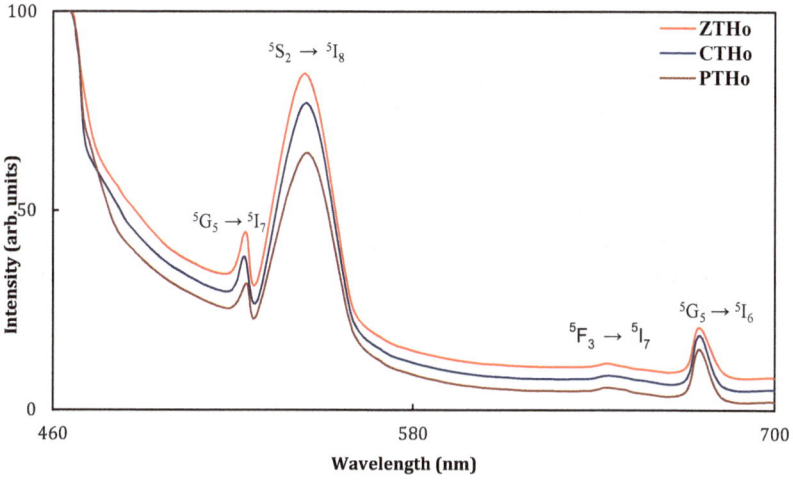

Fig. 2A. Fluorescence spectrum of Ho^{3+} doped ZnF$_2$-MO-TeO$_2$ glass ($\lambda_{exc.}$ = 450 nm)

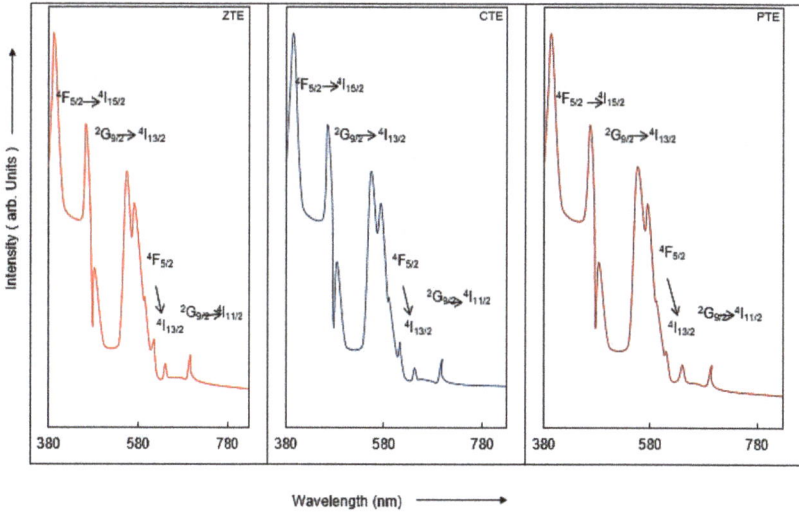

Fig. 2B. Fluorescence spectrum of Er^{3+} doped ZnF$_2$-MO-TeO$_2$ glass ($\lambda_{exc.}$ = 380 nm)

Fig. 3A and Fig. 3B show infrared transmission spectra of crystalline TeO_2 and ZnF_2-MO-TeO_2 glasses containing Ho_2O_3 and Er_2O_3 respectively. The spectra of crystalline TeO_2 has exhibited two fundamental absorption bands at 772 cm^{-1} ($v_1(A_1)$–equatorial band, band 1) and at 650 cm^{-1} ($v_2(A_2)$–axial band, band 2) in the Ho^{3+} doped glasses and the two fundamental absorption bands at 772 cm^{-1} ($v_1(A_1)$–equatorial band, band 1) and at 650 cm^{-1} ($v_2(A_2)$–axial band, band 2) in Er^{3+} doped glasses [33]. However, in the spectra of all the six glasses, the equatorial band is observed to be missing and only the band due to v_{TeO_2ax} vibrations with C_{2v} symmetry is observed. In addition, a band presumably due to MO_nF_m [34] complexes is detected (in the range 1085-1100 cm^{-1}, band 3) in the spectra of all the glasses.

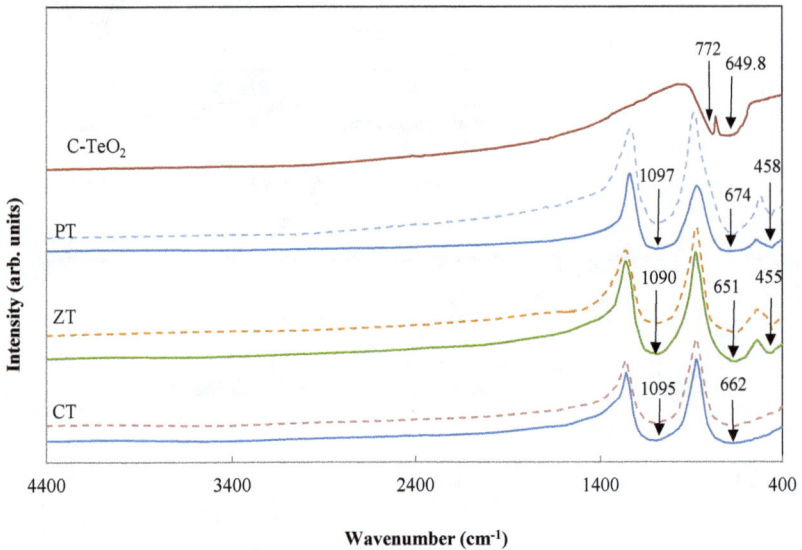

Fig. 3A. Comparison plot showing IR spectra of pure (dotted line) and Ho^{3+} doped ZnF_2-MO-TeO_2 (solid line) glasses

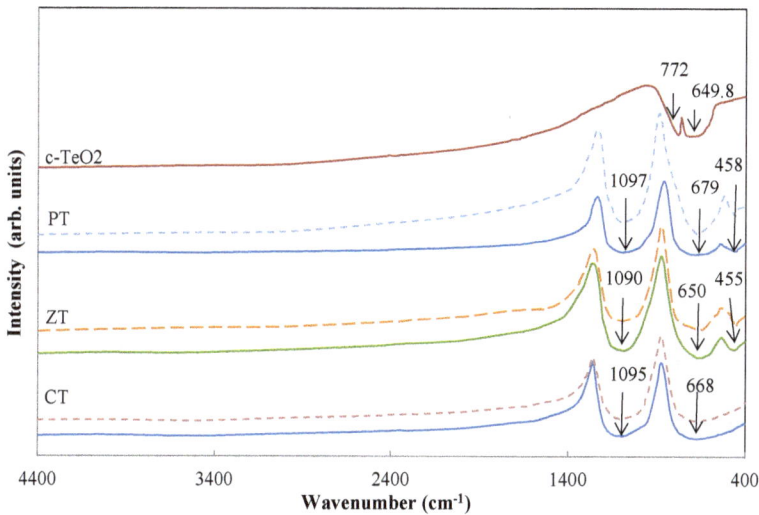

Figure 3B Comparison plot showing IR spectra of pure (dotted line) and Er^{3+} doped ZnF$_2$-MO-TeO$_2$ (solid line) glasses

Further, the IR spectra of glasses containing PbO exhibited absorption bands at 458 cm^{-1} in Ho^{3+} doped glasses and at 460 cm^{-1} in Er^{3+} glasses (assigned to PbO$_4$ structural vibrations) [35]. The IR spectra of glasses containing ZnO exhibited absorption bands at 455 cm^{-1} in Ho^{3+} doped glasses and at 452 cm^{-1} (due to ZnO$_4$ structural vibrations) in Er^{3+} doped glasses [36]. When the glasses are doped with Ho$_2$O$_3$ or Er$_2$O$_3$, the intensity of the axial band is found to decrease with the shifting of meta-centre towards slightly higher wavenumber for all the glasses; however the positions of the other bands remain unchanged. The summary of the data on positions of various bands observed in the IR spectra of ZnF$_2$-MO-TeO$_2$:Ho$_2$O$_3$ and Er$_2$O$_3$ glasses are presented in Table 2.

Table 2 Positions of the absorption bands (in cm^{-1}) of the infrared spectra of Ho^{3+} doped ZnF$_2$-MO-TeO$_2$ glasses

Glass	Band due to ZnO$_4$ and PbO$_4$ units	Band 2 (v_2 – axial band of TeO$_2$)	Band 3 (MO$_n$ F$_m$ complexes)
ZTHo	455	651 (645)*	1090
CTHo	-	662 (660)*	1095
PTHo	458	674 (668)*	1097
ZTE	455	650 (645)*	1090
CTE	-	668 (660)*	1095
PTE	458	679 (668)*	1097
Crystalline- TeO$_2$	649.8 v_2 (axial band of TeO$_2$)	772 v_1 (equatorial band of TeO$_2$)	
* indicate the band position in the spectra of undoped glasses.			

4.   Discussion

ZnF$_2$-MO-TeO$_2$ glasses containing Ho$_2$O$_3$ or Er$_2$O$_3$ have a complex composition and are an admixture of network formers and modifiers. TeO$_2$ belongs to intermediate class of glass forming oxides; it is an incipient glass network former and as such does not readily form glass because the octahedral Te-O polyhedron is highly rigid (when compared with other glass forming oxides like GeO$_2$ etc.,) to get required distortion of Te-O bonds, necessary for forming a stable network. Earlier neutron scattering experiments [37,38] and Raman spectral studies [39,40] on TeO$_2$ glasses containing different modifiers have revealed that the basic building block of TeO$_2$ glass structure is a trigonal bipyramid commonly called TeO$_4$E, where one of the three equatorial directions is occupied by the 5s^2 electronic pair (E) of the tellurium atom with two equatorial bonds of lengths 1.91Å and two axial bonds of lengths 2.08Å [41-43]. The environment of these Te atoms is completed by two other longer interactions of lengths 2.9Å and the three dimensional close packing is constituted from vertices sharing TeO$_4$ groups (Te$_{eq}$-O$_{ax}$-Te) reinforced by weaker Te-O interactions of lengths 2.9Å [44]; this structure leads to long chains of tetrahedrons where the long chain molecules are entwined and the introduction the holmium/erbium ions causes cross linking of the glass structure. If we consider the network modifiers (ZnO, CdO and PbO) to be incorporated between the long chain molecules in the vicinity of holmium/erbium ion, then the symmetry and or covalency of the glass at the holmium/erbium ions should be different for different modifiers. The relatively high intensity of v_{TeO_2ax} -band in the IR spectra of ZTHo/ZTE glass among all the glasses indicates, relatively less degree of disorder of this glass network.

It is well known that there is a shielding of the 4f electrons of the rare-earth ions and this shielding allow these ions to serve as active centers in solid state laser hosts. These ions exhibit sharp absorption and luminescence transitions as surrounding ligand atoms weakly perturb them. The spectral intensities for the observed bands of these glasses that are often expressed in terms of oscillator strength of forced electronic dipole transitions have been analyzed with the help of Judd-Ofelt theory [45] using:

$$f\ (\psi_J, \psi_{J'}) = \frac{8\pi^2 mv}{3h(2J+1)e^2 n_d^2} [X_{ed}\ S_{ed} + X_{md}\ S_{md}] \tag{1}$$

where $(2J+1)$ is the multiplicity of the lower states, m is the mass of the electron and v is the peak absorption in cm^{-1}, $X_{ed} = n_d\ (n_d^2+2)^2/9$ and $X_{md} = n_d^3$. Experimental values of oscillator strengths were evaluated from the expression:

$$f_{exp} = 2.302 \left(\frac{mc^2}{N_A \pi e^2}\right) \int \varepsilon(v)dv \tag{2}$$

where N_A is the Avogadro's number and $\varepsilon(v)$ is the molar absorption coefficient determined from Beer's law. The comparison of experimental and calculated values of oscillator strengths shows a reasonable agreement.

Table 3A The absorption band energies (cm^{-1}), measured (f$_{exp}$ x 10^6) and calculated (f$_{cal}$ x 10^6) oscillator strengths for some transitions of Ho^{3+}:ZnF$_2$-MO-TeO$_2$ glasses.

Transition	ZTHo			CTHo			PTHo		
from $^5I_8$ to	Energy	f_{exp}	f_{cal}	Energy	f_{exp}	f_{cal}	Energy	f_{exp}	f_{cal}
$^5G_5$	23872	2.9	2.04	23866	3.09	1.53	23917	2.2	1.26
$^5G_6$	22093	9.87	11.25	22075	10.7	10.1	22113	7.02	8.54
$^5F_2$	21075	0.48	0.4	21097	0.46	0.37	21057	0.23	0.18
$^5F_3$	20665	0.77	0.71	20652	0.75	0.66	20673	0.39	0.33
$^5F_4$	18611	1.9	1.83	18601	1.91	1.77	18615	1.1	1.1
$^5F_5$	15506	1.8	1.66	15515	1.65	1.62	15496	1.05	1.06
$^5F_6$	8554	0.67	0.64	8560	0.63	0.61	8557	0.33	0.32
$^5F_7$	5115	0.87	0.88	5118	0.82	0.84	5124	0.45	0.45
r.m.s deviation	±0.6119			± 0.5658			± 0.5638		

Table 3B The absorption band energies (cm^{-1}), measured (f$_{exp}$ x 10^6) and calculated (f$_{cal}$ x 10^6) oscillator strengths for some transitions of Er^{3+}:ZnF$_2$-MO-TeO$_2$ glasses.

Transition from $^4I_{15/2}$ to	ZTE			CTE			PTE		
	Energy	f_{exp}	f_{cal}	Energy	f_{exp}	f_{cal}	Energy	f_{exp}	f_{cal}
$^4F_{7/2}$	20492	2.1	1.96	20490	1.82	1.86	20450	1.75	1.75
$^2H_{11/2}$	19231	4.4	4.8	19231	4.1	4.1	19193	4.04	3.5
$^4S_{3/2}$	18348	0.55	0.51	18450	0.52	0.5	18382	0.49	0.47
$^4F_{9/2}$	15358	1.94	1.8	15337	1.82	1.7	15337	1.7	1.6
$^4I_{9/2}$	12450	0.27	0.25	12484	0.26	0.23	12500	0.23	0.21
$^4I_{11/2}$	10280	0.66	0.59	10235	0.56	0.56	10256	0.52	0.52
$^4I_{13/2}$	6532	1.41	1.88	6518.9	1.22	1.70	6531.6	1.1	1.59
r.m.s deviation	± 0.0585			± 0.0302			± 0.0814		

The r.m.s deviations of oscillator strengths of experimental and calculated values of Ho^{3+} ion doped glasses are presented in Table 3A and Er^{3+} ion doped glasses in Table 3B. The relatively small values of these deviations confirm the validity and applicability of Judd-Ofelt theory for the present glasses. The Judd-Ofelt parameters Ω_2, Ω_4 and Ω_6 are computed by the least square fitting analysis of the experimental oscillator strengths using matrix elements [32] and are presented in Table 4.

Table 4 J- O Parameters Ω_λ x10^{20} (cm^{-2}) of Ho^{3+} and Er^{3+} doped ZnF$_2$-MO-TeO$_2$ glasses

Glass	Ω_2	Ω_4	Ω_6	(Ω_4/Ω_6)	Bonding parameter δ'
ZTHo	2.73	1.25	1.03	1.21	0.1469
CTHo	2.38	1.26	0.97	1.30	0.1386
PTHo	2.03	1.03	0.48	2.15	0.0972
ZTE	3.14	1.19	1.43	0.85	-0.3354
CTE	2.39	1.05	1.27	0.87	-0.376
PTE	1.96	0.97	1.20	0.82	-0.3061

The values of Ω_λ show the following order for all the three Ho^{3+} ion doped glasses: $\Omega_2 >$ $\Omega_4 > \Omega_6$. The comparison of J-O parameters shows the highest value of Ω_2 for ZTHo glasses (Table 4) in Ho^{3+} doped glasses. The values of Ω_λ show the following order for all three Er^{3+} ion doped glasses: $\Omega_2 > \Omega_6 > \Omega_4$. The comparison of J-O parameters shows the highest value of Ω_2 for ZTE glasses (Table 4) in Er^{3+} doped glasses.

The bonding parameter (δ'), defined as [46]

$$\delta' = [(1-\bar\beta)/\bar\beta]\times100 \qquad (3)$$

has also been computed for all the three glasses. In eqn. (3), $\bar\beta = \sum_1^N \beta/N$ and β (the nephelauxetic ratio) $= v_c/v_a$. v_c and v_a are the energies in cm^{-1} of the corresponding transitions in the complex and aquo-ion respectively and N refers to the number of levels used to compute $\bar\beta$ values. The computation shows the highest δ' value for ZTHo glass (Table 4) in Ho^{3+} doped glasses and the highest δ' value for PTE glass (Table 4) in Er^{3+} doped glasses.

Table 5A A comparison table of $\Omega_\lambda \times10^{20}$ (cm^{-2}) parameters for a number of other glass systems containing Ho3$^+$ ions.

Glass system	Ref	Ω_2	Ω_4	Ω_6	(Ω_4/Ω_6)	Order
ZTHo	Present work	2.73	1.25	1.03	1.21	$\Omega_2 > \Omega_4 > \Omega_6$
CTHo	Present work	2.38	1.26	0.97	1.30	$\Omega_2 > \Omega_4 > \Omega_6$
PTHo	Present work	2.03	1.03	0.48	2.15	$\Omega_2 > \Omega_4 > \Omega_6$
Lithiumcalcium borate	[47]	6.83	3.15	2.53	1.25	$\Omega_2 > \Omega_4 > \Omega_6$
ZBLA (1)	[48]	2.28	2.08	1.73	1.20	$\Omega_2 > \Omega_4 > \Omega_6$
ZBLA fluoride	[49]	2.43	1.67	1.84	0.91	$\Omega_2 > \Omega_6 > \Omega_4$
Lead silicate	[50]	5.20	1.80	1.20	1.50	$\Omega_2 > \Omega_4 > \Omega_6$
Phosphate	[47]	5.60	2.72	1.87	1.45	$\Omega_2 > \Omega_4 > \Omega_6$
Alumino Bismuth Borate	[21]	5.58	2.59	0.40	6.48	$\Omega_2 > \Omega_4 > \Omega_6$
Tellurite	[47]	6.92	2.81	1.42	1.98	$\Omega_2 > \Omega_4 > \Omega_6$
Aquo ion	[51]	0.36	3.14	3.07	1.02	$\Omega_4 > \Omega_6 > \Omega_2$
Fluorophosphate	[52]	2.08	3.11	1.50	2.07	$\Omega_4 > \Omega_2 > \Omega_6$

Table 5B A comparison table of $\Omega_\lambda \, x10^{20}$ (cm^{-2}) parameters for a number of other glass systems containing Er^{3+} ions.

Glass system	Ref	Ω_2	Ω_4	Ω_6
ZTE	Present work	3.14	1.19	1.43
CTE	Present work	2.39	1.09	1.27
PTE	Present work	1.96	0.97	1.20
Lithium tungsten phosphate	[53]	2.42	1.19	1.45
Lead tungsten phosphate	[53]	3.70	1.43	3.13
CaSiO$_3$–Ca$_3$(PO$_4$)$_2$	[35]	4.50	1.25	0.75
Cadmium borosulphate	[35]	38.96	7.57	11.37
Potassium cadmium borosulphate	[54]	39.62	1.64	10.51
Li$_2$O-SiO$_2$	[55]	4.69	1.29	0.56
Na$_2$O- SiO$_2$	[55]	4.37	0.83	0.27
K$_2$O-SiO$_2$	[55]	5.02	0.71	0.24
Li$_2$O-B$_2$O	[55]	5.81	1.84	1.18
Na$_2$O-B$_2$O$_3$	[55]	6.04	1.59	0.91
K$_2$O-B$_2$O$_3$	[55]	5.38	1.02	0.44
MgSO$_4$+ K$_2$SO$_4$ +ZnSO$_4$	[56]	18.68	2.35	7.15
CaSO$_4$+ K$_2$SO$_4$ +ZnSO$_4$	[56]	15.18	1.65	6.69
BaSO$_4$+ K$_2$SO$_4$ +ZnSO$_4$	[56]	18.37	1.36	9.39
Aquo ion	[46]	1.59	1.95	1.90

In general, the parameter Ω_2 is related to the covalency and structural changes in the vicinity of the the rare-earth ions, in this study Ho^{3+}/Er^{3+} ion (short-range effect) and Ω_4 and Ω_6 are related to the long-range effects. A comparison of J-O parameters, Ω_λ of the present glasses with a number of other glass systems and for aquo-ion is presented in Tables 5A and 5B; the comparison of Ω_2 parameter for the present glasses (Table 5A] &5B) shows the highest value for ZTHo/ZTE glass among all the glasses under investigation, indicating the highest covalent character of this glass. The larger modifier ion (Cd^{2+} ionic radius, 1.03 Å) and also the larger concentration of bonding defects in CTHo glass (evident from the IR spectra) give rise to a larger average distance between the TeO$_4$ chains which results in the average Ho-O/Er-O distance to increase, therefore producing a weaker field around the Ho^{3+}/Er^{3+} ion leading to a low value of Ω_2 when compared with that of ZTHo glasses (Zn^{2+} ionic radius, 0.83 Å). However, in the case of PTHo/PTE glasses the decrease cannot be accounted for, in terms of the ionic radii (Pb^{2+} ionic radius, 1.32 Å) but rather that the Pb^{2+} ion is more strongly bounded to the oxygen that reduces the covalency of the Ho-O/Er-O bond [57]. A similar conclusion can also be drawn from the values of Ω_6, which is related to the rigidity of the host [58]. Further support for this argument can also be cited from the value of the bonding parameter δ'; the value of δ' for these glasses follows the order PTHo < CTHo < ZTHo; indicating the high covalent environment for Ho^{3+} ions in ZTHo glasses. The value of the bonding parameter δ' for Er^{3+} doped glasses follows the order CTE<ZTE indicating the high covalent environment for Er^{3+} ions in ZTE glasses.

Using J-O parameters Ω_λ, the radiative properties of fluorescent transitions from $^5G_5$, $^5S_2$ and $^5F_3$ levels for the present glasses are determined. The spontaneous emission probability A, is calculated using the expression:

$$A(\psi_J, \psi_{J'}) = \frac{64\pi^4 e^2 v^3}{3h(2J'+1)} \left[\frac{n_d(n_d^2+2)^2}{9} S_{ed} + n_d^3 S_{md} \right]$$ (4)

where,

$$S_{ed} = \sum_{\lambda=2}^{6} \Omega_\lambda \left\| U^\lambda \right\|^2$$ (5)

and

$$S_{md} = \frac{e^2 h^2}{16\pi^2 m^2 c^2} (\psi_J \| L + 2S \|^2 \psi_{J'})$$ (6)

In the above equation, (2J'+1) is the multiplicity of the upper state and v is the wavenumber of the fluorescence peak. It may be noted here that the contribution from the magnetic dipoles to the transition probability A for the specific transitions evaluated is found to be either zero or almost negligible.

Then the total emission probability A_T involving all the intermediate terms is calculated using

$$A_T(\psi_J) = \sum A(\psi_J, \psi_{J'})$$ (7)

The radiative lifetime (τ_R) of a state is calculated using the relationship:

$$\tau_R = 1/A_T(\psi_J)$$ (8)

The fluorescent branching ratio is obtained from the equation:

$$\beta_r(\psi_J, \psi_{J'}) = \frac{A(\psi_J, \psi_{J'})}{A_T(\psi_J)}$$ (9)

Finally the stimulated emission cross sections of the measured fluorescent levels are evaluated using:

$$\sigma_P = \frac{A\left(\psi_J,\psi_{J'}\right)\lambda^4}{8\pi c n_d^2 \Delta\lambda} \qquad (10)$$

Where λ is the peak position of the emission line and $\Delta\lambda$ is the effective band width of the emission transitions.

The values of the transitions probability $A(\psi_J, \psi_{J'})$, the total transition probability $A_T(\psi_J)$ and the fluorescence branching ratio β_r evaluated using the equations (7) to (9) for the transitions originated from $^5G_5$ $^5S_2$ and $^5F_3$ levels in Ho^{3+} doped glasses and are presented in Table 6A and for the transitions originated from $^2G_{9/2}$ and $^4F_{5/2}$ levels in Er^{3+} doped glasses are presented in Table 6B.

Table 6A Radiative properties of Ho^{3+} doped ZnF$_2$-MO-TeO$_2$ glasses.

Emission Transition	λ (nm)	$\Delta\lambda$ (nm)	A (s^{-1}) (x10^4)	A_T (s^{-1}) (x10^4)	τ_R (μs)	β_r%	Emission cross Section σ_px10^{20} (cm^2)
ZTHo							
$^5G_5 \rightarrow {}^5I_6$	675	13	0.1001	0.2101	476	47.67	0.85
$^5G_5 \rightarrow {}^5I_7$	525	10	0.1099	0.2101	476	52.32	0.44
$^5S_2 \rightarrow {}^5I_8$	544	32	0.1070	0.1525	655	70.16	0.15
CTHo							
$^5G_5 \rightarrow {}^5I_6$	675	14	0.1001	0.2095	477	47.80	0.79
$^5G_5 \rightarrow {}^5I_7$	524	9.8	0.1093	0.2095	477	52.19	0.45
$^5S_2 \rightarrow {}^5I_8$	545	33	0.1066	0.1522	657	70.00	0.15
PTHo							
$^5G_5 \rightarrow {}^5I_6$	675	14	0.1001	0.2207	481	48.2	0.79
$^5G_5 \rightarrow {}^5I_7$	524	10	0.1076	0.2207	481	51.79	0.433
$^5S_2 \rightarrow {}^5I_8$	545	34	0.1033	0.1490	671	69.32	0.143

The radiative properties of Ho^{3+}/ Er^{3+} ions (or any of Ln^{3+} ions) depend on the number of factors such as network former and modifier of the glass. The parameter β_r (i.e. the branching ratio) of the luminescence transitions characterizes the lasing power of the potential laser transitions. It is well established that an emission level with β_r value nearly equal to 50% becomes a potential laser emission [59]. Referring to the data on emission transitions in the present glass systems, the transition $^5S_2 \rightarrow {}^5I_8$ has the highest value of β_r for all Ho^{3+} doped glasses and the transitions $^2G_{9/2} \rightarrow {}^4I_{13/2}$ and $^4F_{5/2} \rightarrow {}^4I_{15/2}$ have the highest values of β_r in Er^{3+} doped glasses among various transitions; this transition may therefore be considered as possible laser transition. However, the comparison of β_r values of this transition for all the glasses show the largest value for ZTHo/CTE glass indicating these glasses to exhibit better lasing action among the three glasses (Table 6A & 6B).

Table 6B Various radiative properties of Er^{3+} doped ZnF_2-MO-TeO_2 glasses

Emission Transition	λ (nm)	$\Delta\lambda$ (nm)	A (s^{-1})	A_T (s^{-1})	τ_R (μs)	β_r%	Emission cross Section $\sigma_p \times 10^{20}$ (cm^2)
ZTE glass							
$^2G_{9/2} \rightarrow {}^4I_{11/2}$	696	27	252.17	4428.26	0.226	5.69	0.12
$\rightarrow {}^4I_{13/2}$	554	25	1245.08	4428.26	0.226	28.17	0.25
$^4F_{5/2} \rightarrow {}^4I_{13/2}$	643	20	1236.61	3537.46	0.283	34.96	0.56
$\rightarrow {}^4I_{15/2}$	464	18	1838.35	3537.46	0.283	51.97	0.25
CTE glass							
$^2G_{9/2} \rightarrow {}^4I_{11/2}$	700	26	223.93	3892.53	0.257	5.75	0.11
$\rightarrow {}^4I_{13/2}$	552	24	1137.1	3892.53	0.257	29.21	0.23
$^4F_{5/2} \rightarrow {}^4I_{13/2}$	643	20	1117.17	2989.96	0.334	37.36	0.51
$\rightarrow {}^4I_{15/2}$	464	17	1660.79	2989.96	0.334	55.55	0.24
PTE glass							
$^2G_{9/2} \rightarrow {}^4I_{11/2}$	697	26	219.25	3849.8	0.26	5.7	0.1
$\rightarrow {}^4I_{13/2}$	552	25	1099.05	3849.8	0.26	28.55	0.21
$^4F_{5/2} \rightarrow {}^4I_{13/2}$	640	20	1095.05	3043.81	0.329	35.98	0.49
$\rightarrow {}^4I_{15/2}$	461	18	1636.77	3043.81	0.329	53.77	0.21

A comparison of radiative lifetimes for $^2G_{9/2}$ and $^4F_{5/2}$ transitions for a number of other reported glass systems with those of present Er^{3+} doped glasses is given in Table 7 [60-63]. The values of τ_R obtained for these two transitions for the present Er^{3+} doped glasses seemed to be not far away from those of other systems reported.

Table 9 Radiative lifetimes (in µs) of the fluorescent levels $^2G_{9/2}$ and $^4F_{5/2}$ of Er^{3+} in various glass systems

System	Ref	Level $^2G_{9/2}$	Level $^4F_{5/2}$
ZTE glass	present work	226	283
CTE glass	present work	257	334
PTE glass	present work	260	329
Lead tungsten phosphate	[53]	270	340
Calcium tungsten phosphate	[53]	146	154
Zinc sodium sulphate	[60]	169	258
Lithium sodium sulphate	[60]	193	219
Calcium sodium sulphate	[60]	155	170
Potassium sodium sulphate	[60]	206	205
Magnesium sodium sulphate	[60]	228	231
Zinc sodium phosphate	[61]	101	103
Potassium sodium phosphate	[62]	97	115
Sodium aceto phosphate	[63]	191	192

5.   Conclusions

The summary of conclusions drawn from the study of various properties of ZnF_2-MO-TeO_2 glasses doped with Ho_2O_3 and Er_2O_3 ions is as follows: The IR spectral studies indicate relatively less disorder in ZTHo glass network in Ho^{3+} doped glasses and ZTE glass network in Er^{3+} doped glasses. The Judd-Ofelt theory could successfully be applied to characterize the optical absorption spectrum of all the glasses; out of the three J-O parameters Ω_λ, the value of Ω_2, which is related to the structural changes in the vicinity of the Ho^{3+} ion indicates the highest covalent environment for it in ZTHo glasses. The value of Ω_2 of the Er^{3+} ion indicates the highest covalent environment of Er^{3+} ion in ZTE glasses. The radiative transition probabilities evaluated for various luminescent transitions observed in the luminescence spectra of all the Ho^{3+} doped glasses suggest the highest value for $^5S_2 \rightarrow {}^5I_8$ transition (among various transitions originating from $^5G_5$ $^5S_2$ and $^5F_3$ levels) in ZTHo glass. From the evaluated radiative transition probabilities for various luminescent transitions observed in the luminescence spectra of all the Er^{3+} doped glasses suggest the highest value is for $^4F_{5/2} \rightarrow {}^4I_{15/2}$ transition (among various transitions originating from $^2G_{9/2}$ and $^4F_{5/2}$ level) in CTE glass.

6. Acknowledgements

The authors wish to thank Elsevier for permitting to use and reproduce some parts of the chapter from [64], © 2014 Elsevier.

References

[1]   E.F. Chillcce, I.O. Mazali, O.L. Alves, L.C. Barbosa, Opt. Mat. 33, (2011) 389-396. http://dx.doi.org/10.1016/j.optmat.2010.09.027

[2]   H. Desirena, A. Schülzgen, S. Sabet, G. Ramos-Ortiz, E. de la Rosa, N. Peyghambarian, Opt. Mat. 31 (2009) 784-789. http://dx.doi.org/10.1016/j.optmat.2008.08.005

[3]   Tomokatsu Hayakawa, Masahiko Hayakawa, Masayuki Nogami, Philippe Thomas, Opt. Mat. 32 (2010) 448-455. http://dx.doi.org/10.1016/j.optmat.2009.10.006

[4]   G. El-Damrawi, S. Abd-El-Maksoud, Phys. Chem. Glasses 41 (2000) 6-9.

[5]   D. Lezal, J. Pedlikova, P. Kostka, J. Bludska, M. Poulain, J. Zavadil, J. Non-Cryst. Solids 284 (2001) 288-295. http://dx.doi.org/10.1016/S0022-3093(01)00425-2

[6]   M. A. Sidkey, R.A. El-Mallawany, A.A. Abousehly, Y.B. Saddeek, Mat. Chem. Phys. 74 (2002) 222-229. http://dx.doi.org/10.1016/S0254-0584(01)00466-7

[7]   Arshpreet Kaur, Atul Khanna, Carmen Pesquera, Fernando González, Vasant Sathe, J. Non-Cryst. Solids 356 (2010) 864-872. http://dx.doi.org/10.1016/j.jnoncrysol.2010.01.005

[8]   B.V. R. Chowdari, P. Pramoda Kumari, J. Phys. Chem. Solids 58 (1997) 515-525. http://dx.doi.org/10.1016/S0022-3697(96)00160-6

[9]   N. Narasimha Rao, I.V. Kityk, V. Ravi Kumar, P. Raghava Rao, B.V. Raghavaiah, P. Czaja, P. Rakus, N. Veeraiah, J. Non-Cryst. Solids 358 (2012) 702-710. http://dx.doi.org/10.1016/j.jnoncrysol.2011.11.019

[10] Y. Gandhi, I.V. Kityk, M.G. Brik, P. Raghava Rao, N. Veeraiah, J. Alloy. Compd. 508 (2010) 278-291. http://dx.doi.org/10.1016/j.jallcom.2010.08.137

[11] N. Narasimha Rao, I.V. Kityk, V. Ravi Kumar, P. Raghava Rao, B.V. Raghavaiah, P. Czaja, P. Rakus, N. Veeraiah, J. Non-Cryst. Solids 358 (2012) 702-710. http://dx.doi.org/10.1016/j.jnoncrysol.2011.11.019

[12] L. Pavić, N. Narasimha Rao, A. Moguš-Milanković, A. Šantić, V. Ravi Kumar, M. Piasecki, I.V. Kityk, N. Veeraiah, Cer. Inter. 40 (2014) 5989-5996. http://dx.doi.org/10.1016/j.ceramint.2013.11.047

[13] N. Narasimha Rao, I.V. Kityk, V. Ravi Kumar, Ch. Srinivasa Rao, M. Piasecki, P. Bragiel, N. Veeraiah, Cer. Inter. 38 (2012) 2551-2562. http://dx.doi.org/10.1016/j.ceramint.2011.11.026

[14] Y. Gandhi, N. Krishna Mohan, N. Veeraiah, J. Non-Cryst. Solids 357 (2011) 1193-1202. http://dx.doi.org/10.1016/j.jnoncrysol.2010.11.016

[15] M. M. Ahmed, C.A. Hogarth, M.N. Khan, J. Mater. Sci. 19 (1984) 4041-4044. http://dx.doi.org/10.1007/BF00980769

[16] J. D. Lee, Concise Inorganic Chemistry, Oxford: Blackwell Science, 1996.

[17] M. R. Reddy, S.B. Raju, N. Veeraiah, J. Phys. Chem. Solids 61 (2000) 1567-1571. http://dx.doi.org/10.1016/S0022-3697(00)00035-4

[18] B. B. Laud Lasers and Non-Linear optics New Age International (P) Ltd.,(1996)

[19] K. Bhargavi, M. Sundara Rao, V. Sudarsan, Ch. Srinivasa Rao, M. Piasecki, I.V. Kityk, M. Srinivasa Reddy, N. Veeraiah, Opt. Mat. , 36 (2014) 1189-1196. http://dx.doi.org/10.1016/j.optmat.2014.02.027

[20] Ch. Srinivasa Rao, K. Upendra Kumar, P. Babu, C.K. Jayasankar, Opt. Mat. 35 (2012) 102-107. http://dx.doi.org/10.1016/j.optmat.2012.07.023

[21] Sk. Mahamuda, K. Swapna, P. Packiyaraj, A. Srinivasa Rao, G. Vijaya Prakash, Opt. Mat. 36 (2013) 362-371. http://dx.doi.org/10.1016/j.optmat.2013.09.023

[22] T. Satyanarayana, T. Kalpana, V. Ravi Kumar, N. Veeraiah, J. Lumin. 130 (2010) 498-506. http://dx.doi.org/10.1016/j.jlumin.2009.10.021

[23] Z. Yao, Y. Ding, T. Nanba and Y.Miura, Phys. Chem. Glasses 40 (1999) 179.

[24] Shibin Jiang, Michael Myers, Nasser Peyghambarian, J. Non-Cryst. Solids 239 (1998) 143. http://dx.doi.org/10.1016/S0022-3093(98)00757-1

[25] J.A. Pardo, J.J. Pena, R.I. Merino, R. Cases, A. Larrea, V.M. Orera, J. Non-Cryst. Solids 298 (2002) 23. http://dx.doi.org/10.1016/S0022-3093(01)01043-2

[26] Hiroyuki Inoue, Kohei Soga, Akio Makishima, J.Non-Cryst. Solids 298 (2002)270. http://dx.doi.org/10.1016/S0022-3093(01)01052-3

[27] M. Rami Reddy, S. Bangaru Raju, N. Veeraiah, Ind. J. Pure and Appl. Phys 38 (2000) 589.

[28] Shibin Jiang, Tao Luo, Nasser Peyghambarian, J.Non-Cryst.Solids 263 & 264 (2000) 364. http://dx.doi.org/10.1016/S0022-3093(99)00646-8

[29] B.B. Laud, Lasers and Non-Linear optics New Age International (P) Ltd., (1996).

[30] E.P. Chicklis et all App. Phys. Lett , 19 (1971)119.

[31] A. Erbit and H.P. Jenssen, App.Optics 19 (1980) 1729. http://dx.doi.org/10.1364/AO.19.001729

[32] W. T. Carnall, P. R. Fields, K. Rajnak, J. Chem. Phys. 49 (1968) 4424. http://dx.doi.org/10.1063/1.1669893

[33] K. Hirao, S. Todoroki, N. Soga, J. Non-Cryst. Solids 175 (1994) 263-269. http://dx.doi.org/10.1016/0022-3093(94)90019-1

[34] D.K. Durga, P. Yadagiri Reddy, N. Veeraiah, J. Lumin 53 (2002) 99.

[35] G. Srinivasarao, N. Veeraiah, Eur. Phys. J. AP 16 (2001) 11-22. http://dx.doi.org/10.1051/epjap:2001188

[36] P. Subbalakshmi, N. Veeraiah, Ind. J. Eng. Mater. Sci 8 (2001) 275-284.

[37] N. Mochida, K. Takahashi, K. Nakata, Yogyo. Kyokai. Shi 86 (1978) 316-326. http://dx.doi.org/10.2109/jcersj1950.86.995_316

[38] G. A. Clare, C.A. Wright, N.R. Sinclair, F.L. Galeener, E.A. Geissberger, J. Non-Cryst. Solids 111 (1989) 123-138. http://dx.doi.org/10.1016/0022-3093(89)90274-3

[39] S. Neov, V. Kozhukharov, I. Gerasimova, K. Krezhov, B. Sidzhimov, J. Phys. C 2 (1979) 2475-2485. http://dx.doi.org/10.1088/0022-3719/12/13/012

[40] Berthereau, Y. LE Luyer, R. Olazcuaga, Mater. Res. Bul. 29 (1994) 933-941. http://dx.doi.org/10.1016/0025-5408(94)90053-1

[41] T. Sekiya, N. Mochida, A. Ohtsuka, J. Non-Cryst. Solids 144 (1992) 128-144. http://dx.doi.org/10.1016/S0022-3093(05)80393-X

[42] V. Kozhukharov, H. Burger, S. Neov, B. SIdzhimov, Polyhedron 5 (1986) 771-777. http://dx.doi.org/10.1016/S0277-5387(00)84436-8

[43] O. Lindqvist, Acta. Chem. Scand 22 (1968) 977-982. http://dx.doi.org/10.3891/acta.chem.scand.22-0977

[44] F. Folger, Z. Anorg. Allg. Chem. 411 (1971) 111-117. http://dx.doi.org/10.1002/zaac.19754110204

[45] B. R. Judd, Phys. Rev 127 (1962) 750-761. http://dx.doi.org/10.1103/PhysRev.127.750

[46] D. M. Gruen, C. W. Dekock, R. L. Mcbeth, Adv. Chem. Ser 71 (1967) 102-121. http://dx.doi.org/10.1021/ba-1967-0071.ch008

[47] R. Reisfeld, J. Hormadaly, J. Chem. Phys. 64 (1976) 3207-3212. http://dx.doi.org/10.1063/1.432659

[48] K. Tanimura, M. D. Shinn, W. A. Sibley, M. G. Drexhage, R. N. Brown, Phys. Rev. B 30 (1984) 2429-2437. http://dx.doi.org/10.1103/PhysRevB.30.2429

[49] J. Hormadaly, R. Reisfeld, J. Non-Cryst. Solids 30 (1979) 337-348. http://dx.doi.org/10.1016/0022-3093(79)90171-6

[50] F. Fermi, G. Ingletto, C. Aschieri, M. Bettinelli, Inorgan. Chemica Acta 163 (1989) 123-125. http://dx.doi.org/10.1016/S0020-1693(00)83437-4

[51] J. R. Quagliano, F. S. Richardson, M. F. Reid, J. Alloy. Compd. 180 (1992) 131-139. http://dx.doi.org/10.1016/0925-8388(92)90372-G

[52] Meng Wang, Lixia Yi, Guonian Wang, Hu Lili, Junjie Zhang, Solid State Commun. 149 (2009) 1216–1220. http://dx.doi.org/10.1016/j.ssc.2009.04.021

[53] P. Subbalakshmi, N. Veeraiah, J. Phys. Chem. Solids 64 (2003) 1027. http://dx.doi.org/10.1016/S0022-3697(02)00370-0

[54] C.K. Jayasankar, V.V. Ravi Kanth Kumar, Pramana 48 (1997) 1151. http://dx.doi.org/10.1007/BF02845890

[55] H. Takebe, Y. Nageno, K. Morinaga, J. Am. Ceram. Soc 77 (1994) 2132. http://dx.doi.org/10.1111/j.1151-2916.1994.tb07108.x

[56] Y. Subramanyam, L.R. Moorthy, S.V.J. Lakshman, Mater. Lett 91 (1988) 46.

[57] J.A. Capobianco, P.P. Proulx, M. Bettinelti, F. Negrisolo, Phys. Rev. 42 (1990) 5936-5944. http://dx.doi.org/10.1103/PhysRevB.42.5936

[58] C. K. Jorgensen, R. Reisfeld, J. Less-Common. Met. 93 (1983) 107-112. http://dx.doi.org/10.1016/0022-5088(83)90454-X

[59] J. L. Adam, W. A. Sibley, J. Non-Cryst. Solids 76 (1985) 267-279. http://dx.doi.org/10.1016/0022-3093(85)90004-3

[60] S.V.J. Lakshman, Y.C. Ratnakaram, J. Non-Cryst. Solids 94 (1987) 222 . http://dx.doi.org/10.1016/S0022-3093(87)80292-2

[61] S.V.J. Lakshman, A. Suresh kumar, Phys. Chem. Glasses 29 (1988) 146.

[62] S.V.J. Lakshman, A. Suresh kumar, J. Phys. Chem. Solids 49 (1988) 133. http://dx.doi.org/10.1016/0022-3697(88)90042-X

[63] S.V.J. Lakshman, A. Suresh kumar, Lanthanide. Actinide. Res 2 (1988) 243.

[64] C. Laxmikanth, J. Anjaiah, P. Venkateswara Rao, B. Appa Rao, N. Veeraiah., Luminescence and spectroscopic properties of ZnF2–MO–TeO2 glasses doped with Ho^{3+} ions. Journal of Molecular Structure 1093 (2014)166-171. http://dx.doi.org/10.1016/j.molstruc.2015.03.018

CHAPTER 7

Luminescence and energy transfer phenomena in lanthanide ions doped phosphor and glassy materials

G. Bhaskar Kumar[a], B. Vengla Rao[a], B.Chandra Babu[b], Graham Hungerford[c]
Sooraj H Nandyala[d*] and J.D. Santos[e]

[a]Department of Humanities and Sciences, Sri Venkateswara College of Engineering and Technology, (SVCET) RVS Nagar, Chittoor-517127, Andhra Pradesh, INDIA

[b]Shenzhen Key Laboratory of Advanced Materials, Department of Materials Science and Engineering, Shenzhen Graduate School Harbin Institute of Technology Shenzhen 518055, CHINA

[c] HORIBA Jobin Yvon IBH Ltd., 133 Finnieston Street, Glasgow G3 8HB, UK

[d]School of Metallurgy and Materials, University of Birmingham, Edgbaston, Birmingham B15 2TT, UK

[e]CEMUC, Departamento de Engenharia Metalúrgica e Materiais, Faculdade de Engenharia, Universidade do Porto, Rua Dr Roberto Frias, 4200-465 Porto, PORTUGAL

Abstract

This chapter aims to explain the basic mechanism of phosphor materials and the luminescence behaviour of glasses doped with certain rare earth ions. It will also describe phosphor based white light emitting diodes and their significance. The photoluminescence properties of $Ca_3Y_2Si_3O_{12}$ doped with Ce^{3+}, Tb^{3+} and Tb^{3+} & Ce^{3+} co-doped phosphors prepared by the sol-gel method are presented. Photoluminescence (PL) spectra revealed a brighter and broader violet-blue colour emission from the Ce^{3+} $(5d(^2D)$ $\rightarrow^2F_{5/2,7/2})$ and an intense sharp green emission (545 nm) colour from the Tb^{3+} $(^5D_4\rightarrow^7F_5)$ doped phosphors respectively. For Tb^{3+} and Ce^{3+} co-doped phosphors, a strong green emission $(^5D_4\rightarrow^7F_5)$ has been observed upon excitation with a UV wavelength (242nm). An energy transfer phenomenon from Ce^{3+} to Tb^{3+} was seen in these co-doped nanocrystalline phosphors. A dependency of the PL intensity on the doping concentration of Ce^{3+} was found, for a fixed concentration of Tb^{3+}. The luminescent properties of the Ce^{3+} or Dy^{3+} singly doped and $(Ce^{3+}$ & $Dy^{3+})$ co-doped $Ca_3Y_2Si_3O_{12}$ novel phosphors are also reported. The Ce^{3+} doped phosphor showed a brighter /broader violet-blue colour emission (389nm), which is attributed to the parity and spin allowed 5d–4f transition.

Photoluminescence spectra reveal that the white emission originated from the mixtures of two characteristic luminescence of Dy^{3+} ion, i.e. the $^4F_{9/2}$-$^6H_{15/2}$ at 473 nm blue emission, and $^4F_{9/2}$-$^6H_{13/2}$ at 580 nm yellow emission. The co-doping of Ce^{3+} significantly enhanced the luminescence of Dy^{3+} upon UV excitation (at a wavelength of 242nm) and the optimum co-dopant concentration of Ce^{3+} was found to be 3 mol%. Finally, the visible-NIR luminescence performance of Nd^{3+} and Er^{3+} ions in silver zinc borate host glasses are discussed by means of visible and NIR luminescence spectral profiles.

Keywords

Phosphor Materials, Photoluminescence Studies, Glasses

Contents

1.   State of the art

The development of novel and potential ultraviolet (UV)/blue LEDs based on wideband gap semiconductor such as GaN led to considerable progress in the field of solid state lighting. In 1996, a white light LED that had a combination of a blue GaN LED chip and a yellow phosphor $(Y_{1-x}Gd_x)_3(Al_{1-y}Gd_y)_5O_{12}:Ce^{3+}$ (YAG: Ce) was commercialized. Although white light was easily achieved using the YAG: Ce based system, the individual degradation rate between the blue LED and yellow phosphor caused chromatic aberration and poor white light performance after a long period of use. Therefore, developing a single phased phosphor that can emit white light when excited by the UV LED might be an excellent alternative to replace the YAG: Ce based system. The absorption spectra of Ce^{3+} ions usually consist of broad bands in UV, relating to transitions between the $4f$ and $5d$ energy levels. The efficient energy transfer from Ce^{3+} ions to the rare earth ions results in a stronger luminescence intensity of the rare earth ions in these co-doped materials. Hence, the Ce^{3+} ion acts as an efficient sensitizer for rare earth ions.

For more than a century, sulphide phosphors have been widely studied as luminescent host lattices. However, sulphide phosphors are not stable or bright enough for many applications. So there has been a great demand for new type of host lattice substitute. Over the past decade extensive research has been carried out on oxide-based phosphors, because of their superior colour richness and good chemical and thermal stability compared to non-oxide materials, e.g., sulphides doped with the d- or s-series elements. Alkaline earth silicates are considered as suitable hosts with high chemical stability and water-resistant properties. Many reports on $CaO\text{-}SiO_2$ based systems have been published owing to its promising luminescence behaviour [1-3]. Hence in this chapter it is intended to explain the luminescence and energy transfer phenomena of lanthanide/rare-earth ions doped in $Ca_3Y_2Si_3O_{12}$ phosphor and Nd^{3+} and Er^{3+} ions doped silver zinc borate host glassy materials.

2.   What is phosphor?

A phosphor is usually a powder material that glows when bombarded by a beam of electrons or other high speed sub atomic particles. These phosphors also glow when exposed to ultraviolet radiation, X-rays or γ-rays. Generally phosphors would have either one or more luminescent ions, which are normally transition-metal ions ($3d^n$) or rare-earth ions ($4f^n$) sometimes in the form of dual ions (3d-3d, 4f-4f, 3d-4f) [5,6]. In fact, the word luminescence was first used in 1888 by a German physicist namely Eilhardt Wiedeman for the emission of light by certain materials upon excitation from high energy radiation (X-rays or UV) or electrons. Such an emitted light would normally be pleasant and hence, it has been popularly known as 'cold light' or fluorescence light [5-7].

Mechanism of luminescence in phosphor systems

An inorganic phosphor system consists of two parts;

 i) The host (or inorganic compound)

 ii) The activator (or the added transition/rare-earth metal cation).

The host needs to be transparent (or a non-absorbing cation) to the radiation source used for the "excitation" process and the activator is just required to activate the host material. This combination has several advantages; the most important is that both the type and amount of activator can be precisely controlled to tune the system for the required application.

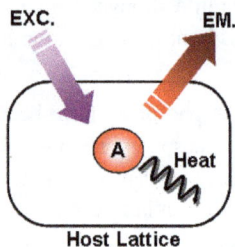

Figure 1. A physical model of luminescence, where Exc is excitation, EM is emission or radiative return to the ground state. Heat signifies a non-radiative return to the ground state.

In the mechanism described in Fig 1 and Fig 2 the host lattice does not participate in the emission phenomenon, but simply tightly contains the activator ion (RE^{3+}) while the activator absorbs the energy and emits the radiation. A well-known example is ruby $Y_2O_3:Eu^{3+}$.

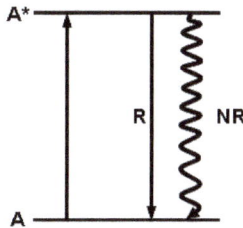

Figure 2. *Energy Level Diagram, where A* is the excited State of activator, A is the ground state of activator, R is the radiative transition to the ground state (emission) and NR relates to a non-radiative transition (or heat).*

However, if another ion is added to the host lattice, this opens the possibility for that ion to absorb the excitation energy and then transfer it to an activator. Such an absorbing ion is known as a sensitizer and the phenomenon is shown in Fig. 3.

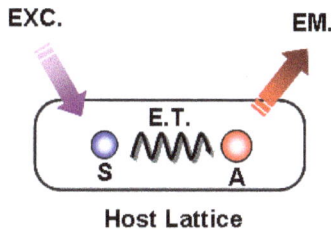

Figure 3. *Energy Transfer from a sensitizer S to an activator A. Energy transfer is indicated by E.T.*

A well-known example for this type of luminescence is $Ca_5 (PO_4)_3F:Sb^{3+}$, Mn^{2+}. Ultraviolet radiation is not absorbed by Mn^{2+}, but only by Sb^{3+}. Under ultraviolet irradiation, the emission consists partly of blue Sb^{3+} emission and partly of yellow Mn^{2+} emission. Since the Mn^{2+} ion was not excited directly; the excitation energy was transferred from Sb^{3+} to Mn^{2+}. Another mechanism of luminescence is one in which the host lattice is excited and then transfers its excitation energy to the activator. A familiar example of this type of system is $CaWO_4$, where the tungstate group (WO_4^{2-}) is the luminescent centre and $CaNb_2O_6$, where the niobate octahedron ($Nb_2O_6^{2-}$) is the luminescent centre. This class of phosphors is known as "self-activated" because the host lattice both absorbs and emits the radiation [5-7].

3.   Energy transfer mechanism

Energy transfer can occur between a pair of dissimilar luminescent centres or between identical luminescent centres. The case of dissimilar luminescent centres involves two centres: a sensitzer S and an activator A separated in a solid by a distance R, where R is assumed to be so short that S and A have a non-vanishing interaction with one another. This means that if S is in an excited state while A is in the ground state, then the relaxed excited state of S may transfer its energy to A. In this case, resonance has to be satisfied, that is, the energy differences between the ground and excited states have to be equal and a suitable interaction has to exist for energy transfer to occur. This interaction can be an exchange interaction where there is a wave function overlap or an electric or magnetic multipolar interaction.

Energy transfer between identical luminescent centres results in the phenomenon of concentration quenching of luminescence and a weak-coupling scheme or strong coupling scheme can be applied. A typical case of stronger coupling is well-known in groups like tungstates and vanadates, which are oxidic anions with a central metal ion which has no d electrons, e.g. WO_4^{2-}, WO_6^{6-}, VO_4^{3-} and MoO_4^{2-}. The nature of the luminescent species determines the strength of the electron lattice coupling, for example, in the weak coupling case, the zero phonon line dominates and narrow peaks, as in Eu^{3+} emission, are observed. In the excited state, the electronic charge has moved from the oxygen ligands to the central metal ion and this is considered to be a charge transfer state. Actually, the amount of charge transfer is normally small but a significant amount of electronic reorganization occurs whereby electrons are promoted from bonding orbitals in the ground state to antibonding orbitals in the excited state. This leads to a large value of ΔR in the configurational coordinate diagram, and broad spectral bands [5-7].

4.   Charge transfer transitions

Jorgensen was the first to assign the broad and strong absorption bands in the spectra of the trivalent lanthanides to either charge-transfer (CT) or $4f \rightarrow 5d$ transitions. As a general rule the CT bands shift to lower energies with increasing oxidation state, whereas Rydberg transitions (such as $4f \rightarrow 5d$ transitions) shift to higher energies. It may, therefore, be expected that the lowest absorption bands of the tetravalent ions will relate to CT transitions and those of the divalent lanthanide ions to $4f \rightarrow 5d$ transitions. In the case of trivalent lanthanide ions, depending upon the number of f-electrons in the ground state, the first allowed transition may be either a CT transition of a $4f \rightarrow 5d$ transition. The stability of the half-filled and completely filled shells serves as a starting point to predict which of the two transitions is to be expected. The charge transfer transitions are at relatively low energy in the case of Eu^{3+} ($4f^6$), Sm^{3+} ($4f$) and Yb^{3+} ($4f^{13}$) and the $4f \rightarrow 5d$

transitions are at relatively low energy in the case of Ce^{3+} $(4f^1)$, Pr^{3+} $(4f^2)$ and Tb $(4f^8)$. Charge transfer (CT) transitions are transitions involving ligands and/or central ions, while the charge configuration is changed [4-6].

5.   Prominent emission colour of rare earth ions

Ions with no 4f electrons, i.e., Sc^{3+}, Y^{3+}, La^{3+}, and Lu^{3+}, have no electronic energy levels that can induce excitation and luminescence processes in or near the visible region. In contrast, the ions from Ce^{3+} to Yb^{3+}, which have partially filled 4f orbitals, have energy levels characteristic of each ion and show a variety of luminescence properties around the visible region. Many of these ions can be used as luminescent ions in phosphors, mostly by replacing Y^{3+}, Gd^{3+}, La^{3+}, and Lu^{3+} in various compound crystals. The lowest excited 4f level of Gd^{3+} $(^6P_{7/2})$ gives rise to sharp-line luminescence at ~315 nm and can sensitize the luminescence of other rare-earth ions. The energy levels of the CT state and the $4f^65d^1$ states are the highest among rare-earth ions, so that Gd^{3+} causes no quenching in other rare-earth ions. As a consequence, Gd^{3+} serves, as Y^{3+} does, as a good constituent cation in host crystals to be substituted by luminescent rare-earth ions. The four lower-lying levels of Nd^{3+} provide a condition favourable to the formation of population inversion. For this reason, Nd^{3+} is used as the active ion in many high-power, solid-state lasers (at 1.06 μm wavelength).

Table 1. *Prominent Emission Colour of Rare Earth Ions*

Rare Earth Ion	Emission Transition	Emission Wavelength (nm)	Emission Colour
Eu^{3+}	$^5D_0 \rightarrow {}^7F_2$	610–630	Red
Eu^{2+}	Wavelength positions of the emission bands depend very much on hosts		Varies from near-UV - Red
Tb^{3+}	$^5D_4 \rightarrow {}^7F_5$	535-550	Green
Sm^{3+}	$(^4G_{5/2} \rightarrow {}^6H_{7/2})$	600-610	Reddish Orange
Sm^{2+}	$^5D_0 \rightarrow {}^7F_1$	696	Red
Dy^{3+}	$^4F_{9/2} \rightarrow {}^6H_{15/2}$	470-500	White
	$^6F_{15/2} \rightarrow {}^6F_{11/2}$	570 -600	
Dy^{4+}	$^5D_4 \rightarrow {}^7F_3$	630	Red
	$^5D_4 \rightarrow {}^7F_5$	525	Green
Pr^{3+}	$^1D_2 \rightarrow {}^3H_4$	600-610	Red
Ho^{3+}	$^5S_2 \rightarrow {}^5I_8$	525-540	Green
Tm^{3+}	$^1G_4 \rightarrow {}^3H_6$	460-480	Blue
Er^{3+}	$^4S_{3/2} \rightarrow {}^4I_{15/2}$	535-550	Green
Ce^{3+}	Wavelength positions of the emission		Varies from near -

	bands depend very much on hosts		ultraviolet to green
Yb^{2+}	$4f^{14} \leftrightarrow 4f^{13}5d^1$	432	Blue

Er^{3+} ions embedded in an optical fibre function as an optical amplifier for 1.55 μm semiconductor laser light. Population inversion is realized between lower sublevels of $^4I_{13/2}$ and upper sublevels of $^4I_{15/2}$. This technology has been developed for optical amplification in the long-distance optical fibre communication systems [1-6]. Other rare-earth ions suitable for the fibre-optic applications include Pr^{3+}, Dy^{3+} and Nd^{3+} at 1.3 μm; Tm^{3+} at 1.2 and 1.4 μm.

6. Applications of luminescent materials

Emissive materials, particularly phosphors, are used in most display technologies, including electroluminescent, cathode ray tube, field emission, plasma, and liquid crystal as direct light emitters or as illumination sources. The performance of display products depends upon phosphor efficiencies, spectral distribution, long-term stability and electrical characteristics. Often it is the constraints of the emissive materials that limit improvements in the display properties. Fig. 4 gives the summary of phosphor devices and its applications [4].

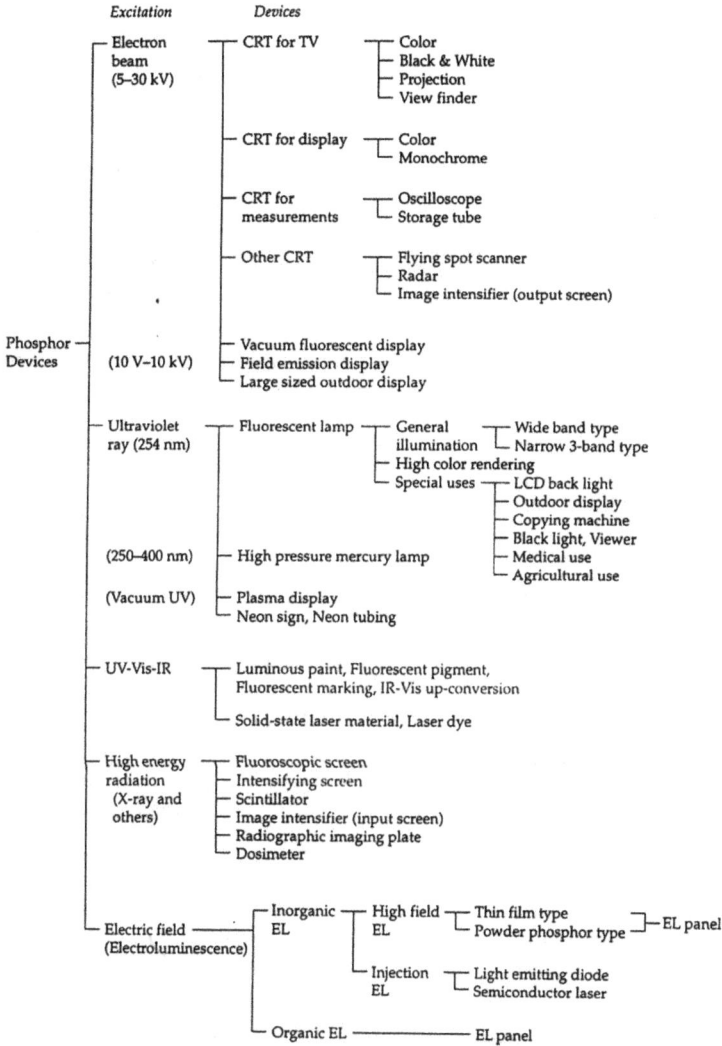

Figure 4. *Phosphor Devices and its applications [4].*

7.   Evolution of lighting

Since the realization of GaN-based LEDs, more and more interest has been focused on WLEDs. Over the incandescent and fluorescent lamps, WLEDs have many advantages such as; long lifetime, energy saving, safety, reliability, maintenance and environmentally friendly characteristics. One way of producing white light (see Fig. 5) is by the combination of a blue LED with a yellow phosphor, blending the blue light from LED and yellow light from the phosphor results in white light. However, warm white-light illumination usually cannot be achieved by this approach because of the red deficiency of the spectral emission in this system. This drawback has limited the possible applications of WLEDs in the medical and architectural lighting fields. Another approach that might solve this problem is by using an ultraviolet (UV) LED chip coated with three emitting blue, green, and red phosphors to generate a warm white light. This approach provides WLEDs with excellent colour rendering indexes.

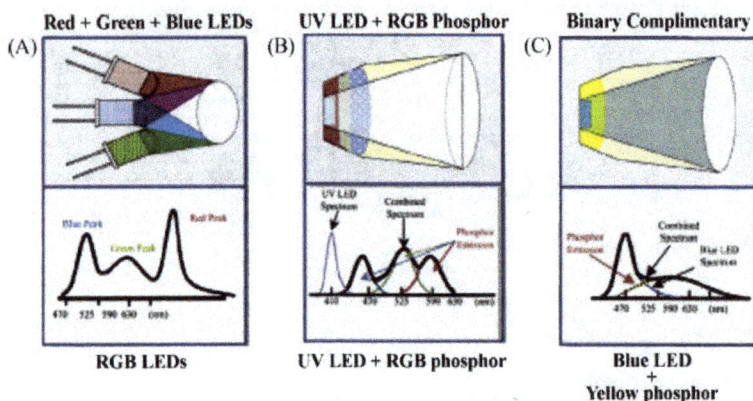

Figure 5. *Three popular methods of generating white light [5].*

However, in this WLED system, the blue emission efficiency is poor because of the strong reabsorption of the blue light by the green and red phosphors. Compared to WLEDs with multi-component emitters, WLEDs based on a single component emitter can overcome these disadvantages and also has other advantages like improved stability, better reproducibility and a simpler fabrication process. In addition, a device consisting of white LEDs with multiple emitting components is very complicated and difficult to be realized. So it is necessary to develop a emission tunable white light emitting phosphor which can avoid these problems [2,3].

Dramatic changes are unfolding in lighting technology (see Figure 6). Semiconductor light emitting diodes have already begun to displace incandescent bulbs in many applications, particularly those requiring durability, compactness, cool operation and/or directionality (e.g., traffic, automotive, display and architectural/directed area lighting. Further major improvements in this technology are believed achievable. The luminous efficiency of LED systems with 80 – 100 lm/W is currently available. If a luminous efficiency of LED systems with 200 lm/W is achieved in the visible, the result would be the holy grail of lighting. A 200 lm/W white light source is two times more efficient than fluorescent lamps and ten times more efficient than incandescent lamps. This new white light source would change the way we live and the way we consume energy. It has the potential of the following cumulative efforts by the year 2020:

1. Decreasing by 50% the amount of electricity used for lighting.

2. Decreasing by 11% the total consumption of electricity.

3. Producing a reduction of 133 GW of electricity.

4. Freeing over 133 GW of electric generating capacity for other uses.

5. Decreasing by 50% the amount of electricity used for lighting.

6. Over the same time period, reducing carbon emissions by 259M tons due to lower demands on electricity.

These spectacular benefits constitute the promise of semiconductor lighting envisioned for the year 2020. Hence many countries all over the world propose federally-funded industry driven R&D effort involving national laboratories and universities to accelerate the development of this "semiconductor lighting revolution" [1, 8-10].

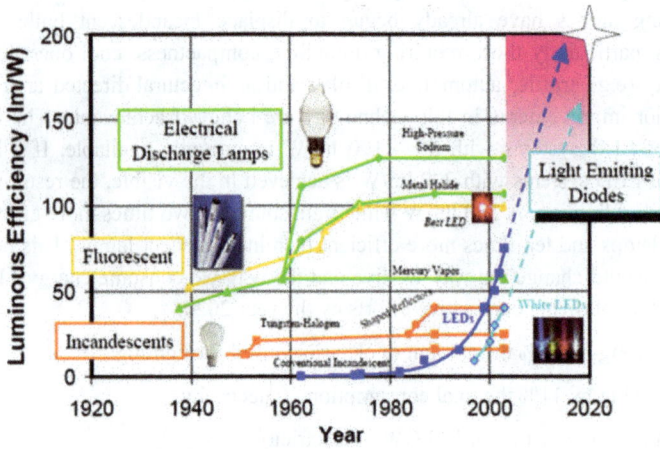

Figure 6. Evolution of Light [8].

8. CIE Chromaticity Diagram

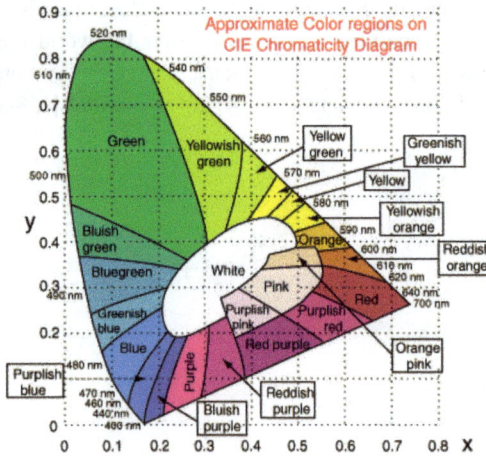

Figure 7. CIE chromaticity diagram.

The CIE 1931 colour spaces were the first defined quantitative links between physical pure colours (i.e. wavelengths) in the electromagnetic visible spectrum, and physiological perceived colours in human colour vision. The mathematical relationships that define these colour spaces are essential tools for colour management; important when dealing with coloured inks, illuminated displays and recording devices such as digital cameras. The CIE 1931 RGB colour space and CIE 1931 XYZ colour space were created by the International Commission on Illumination (CIE) in 1931. They resulted from a series of experiments done in the late 1920s by William David Wright and John Guild. The experimental results were combined into the specification of the CIE RGB colour space, from which the CIE XYZ colour space was derived. The diagram given here (Fig 7) is associated with the 1931 CIE standard. Revisions were made in 1960 and 1976, but the 1931 version remains the most widely used version. The diagram (Fig. 7) is a rough rendering of the 1931 CIE colours on the chromaticity diagram. The table 2 shows the colours which can be matched by combining a given set of three primary colours (such as the blue, green, and red of a colour television screen) are represented on the chromaticity diagram by a triangle joining the coordinates for the three colours [11].

9.   Plotting colours in CIE diagram

In general the photoluminescence emission spectra of the phosphors with 1nm wavelength separation is recorded and the emission spectral values corresponding to each colour should be copied and pasted in the CIE excel sheet. By doing so we can get the x, y, z co-ordinates of the CIE diagram. By inputting those values in the CIE software we can get the CIE diagram for the prepared phosphor [12].

10.   Experimental

Phosphors prepared by a solid state reaction method has some disadvantages, such as high sintering temperature, inhomogeneous mixing and irregularly shaped and aggregated particles in the bulk form. Wet chemical methods, e.g. sol-gel, can overcome these disadvantages. The main differences of wet chemical over solid state methods are much smaller grains and, usually, lower temperature and shorter duration. The sol-gel process is a method for obtaining materials from solutions, in which the gel formation is present at one of the process stages. A crucial role in the sol-gel process is played by the process of solvent removal from the gel. Most products of sol-gel synthesis are used as precursors in obtaining nanostructures. In the Pechini Method, metal salts or alkoxides are introduced into a citric acid solution with ethylene glycol. The formation of citric complexes is believed to balance the difference in individual behaviour of ions in solution, which results in a better distribution of ions and prevents the separation of components at later process stages. The polycondensation of ethylene glycol and citric acid starts above 100°C, resulting in polymer citrate gel formation. When the heating temperature exceeds 400°C, oxidation and pyrolysis of the polymer matrix begin, which lead to the formation of amorphous oxide/carbonate precursor. Further heating of this precursor results in the formation of the required material with a high degree of homogeneity and dispersion.

Table 2. *Colour co-ordinates of CIE Diagram*

Colour name	Red	Green	Blue
Red	191	27	75
Pink	245	220	208
Reddish orange	216	119	51
Orange pink	240	204	162
Orange	228	184	29
Yellowish orange	231	224	0
Yellow	234	231	94
Greenish yellow	235	233	0
Yellow green	185	214	4
Yellowish green	170	209	60
Green	0	163	71
Bluish green	24	162	121
Bluegreen	95	164	190
Greenish blue	110	175	199
Blue	92	138	202
Purplish blue	88	121	191
Bluish purple	92	102	177
Purple	246	85	158
Reddish purple	196	64	143
Purplish pink	243	208	219
Red purple	175	35	132
Purplish red	209	65	136
White	255	255	255

$Ca_3Y_2Si_3O_{12}$:RE^{3+} (Ce^{3+} and/or Dy^{3+}, $Tb3^+$) phosphors were prepared by adapting sol-gel method. Stoichiometric amounts of $CaCO_3$, $Y(NO_3)_3$, $Ce(NO_3)_3$, $Tb(NO_3)_3$, $Dy(NO_3)_3$ and Tetraethyl orthosilicate (TEOS) were used as starting materials. $CaCO_3$ and $RE(NO_3)_3$ were dissolved in HNO_3 and DI water respectively. The TEOS dissolved into a double distilled water and ethanol (volume ratio = 1:4) solution, containing citric acid as a chelating agent for the metal ions. The molar ratio of metal ions to citric acid was 1:2. The above two solutions were mixed and stirred for 2h at room temperature to form a homogeneous solution. Ammonia was added to the above solution to neutralize the excess nitrate and the gel thus formed was dried in ambient atmosphere at 100°C. The dried gel was calcined at 400°C for 3h. The precursor thus obtained was again calcined at 1000°C in air to obtain the final product.

10.1   Luminescence and energy transfer of Ce^{3+} and/or Tb^{3+} in $Ca_3Y_2Si_3O_{12}$

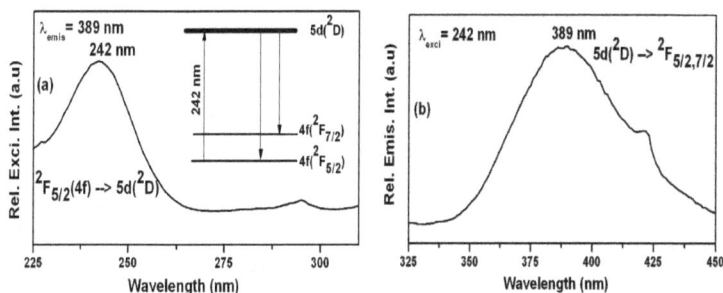

Figure 8. (a) Excitation and (b) emission spectrum of $Ca_3Y_{1.95}Si_3O_{12}$: $Ce_{0.05}$ phosphors.

The excitation spectrum of $Ca_3Y_{1.95}Si_3O_{12}$:$Ce_{0.05}$ phosphor is shown in Fig 8(a) monitored with an emission wavelength 389 nm. The excitation spectrum shows a band at 242 nm along with a weak band at 295 nm, which is assigned to the electronic transition of Ce^{3+} ion ($^2F_{5/2}(4f) \rightarrow 5d(^2D)$). Fig. 8(b) shows the emission spectrum of $Ca_3Y_{1.95}Si_3O_{12}$: $Ce_{0.05}$ phosphor upon excitation with 242 nm. The emission of Ce^{3+} includes a broad band centred at 389 nm and a shoulder at 422 nm, which are assigned to the parity allowed transitions of the lowest component of the ^2D state to the spin-orbit components of the ground state, $^2F_{5/2}$ and $^2F_{7/2}$ of Ce^{3+} respectively. The energy difference between the two peaks is 2010 cm^{-1}, basically agreeing with the ground state splitting of Ce^{3+}. It is well known that the energy difference between the $^2F_{5/2}$ and $^2F_{7/2}$ doublets in the $4f^1$ configuration of the Ce^{3+} ion [13-15].

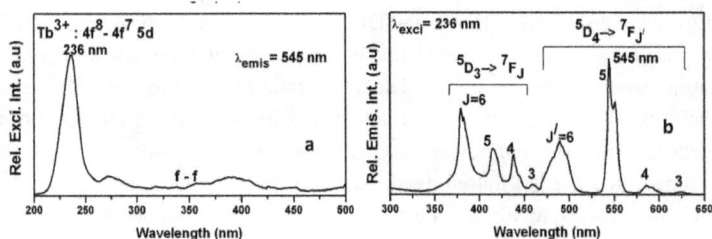

Figure 9. (a) Excitation and (b) emission spectrum of $Ca_3Y_{1.95}Si_3O_{12}$ doped with $Tb_{0.05}$ nano crystalline phosphor.

Fig. 9(a) shows the excitation and emission spectrum of $Ca_3Y_{1.95}Si_3O_{12}$ doped with $Tb_{0.05}$ nano crystalline phosphor. For Tb^{3+} ions with $4f^8$ electrons configuration, the ground states are ^7F_J. When one electron is promoted to 5d shell, it gives rise to two $4f^75d^1$ excitation states: the high-spin state with ^9D_J configurations or low-spin state with ^7D_J configurations. Obviously, ^9D_J states will be lower in energy according to Hund's rule, and the transitions between ^7F_J and ^7D_J are spin-allowed, while the transitions between ^7F_J and ^9D_J are spin-forbidden. Therefore, the Tb^{3+} ion in a specific host exhibits two groups of f–d transitions: the spin-allowed f–d transitions are strong, with higher energy; the spin-forbidden f–d transitions are weak, with lower energy. In the excitation spectrum (Fig. 9(a)), the strong band at 236 nm and weak band at 274 nm are observed to be the spin-allowed and spin-forbidden f–d transitions ($4f^8 \rightarrow 4f^75d^1$), respectively. As the energy difference between the bands at 236 and 274 nm is about 5870 cm^{-1}, it is near the average energy difference ΔETb(sa),Tb(sf) between the spin-allowed f-d transition and the spin-forbidden transition for Tb^{3+}(6000 cm^{-1}) [16]. A series of relatively weak peaks in the range 300-400 nm are ascribed to the transitions within $4f^8$ configuration of Tb^{3+} ion. Effective excitation in the spin allowed $4f^8 \rightarrow 4f^75d^1$ transition is possible in view of its high intensity.

The Tb^{3+} ion can be used as an activator in blue and green phosphors. At lower concentrations it mainly emits blue light, but at higher concentrations it emits green light only. Fig. 9(b) gives the emission spectra of $Ca_3Y_2Si_3O_{12}$:Tb^{3+} excited at 236nm. At this Tb^{3+} concentration, both blue emission from the $^5D_3$ levels and green emission from $^5D_4$ levels have been observed. From the transitions of $^5D_3 \rightarrow ^7F_5$ (415 nm) and $^5D_4 \rightarrow ^7F_5$ (545 nm), the energy difference between $^5D_3$ and $^5D_4$ can be calculated, which amounts to 5748 cm^{-1}. According to Schuurmans and Van Dijk and Blasse, it is found that the radiative rate is approximately equal to the non-radiative rate if the energy gap in the non-radiative transition equals five times the maximum phonon frequency [17]. In $Ca_3Y_2Si_3O_{12}$ this situation occurs because the maximum frequency of the phonons in the silicate lattice is about 950 cm^{-1} and the energy gap between the $^5D_3$ and $^5D_4$ levels amounts to 5763 cm^{-1}, as calculated above. Indeed, the observed amounts of the $^5D_3$ and $^5D_4$ emissions are of the same order of magnitude.

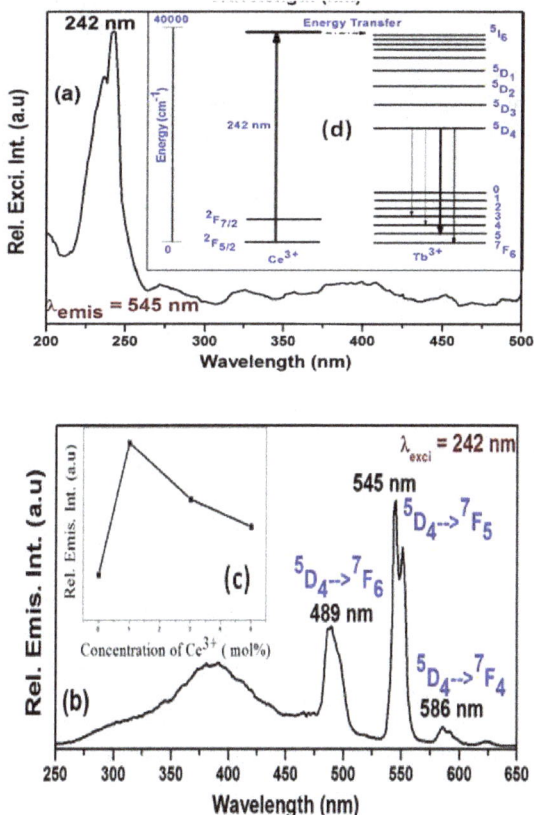

Figure 10. (a) Excitation (b) Emission spectrum of $Ca_3Y_{1.94}Si_3O_{12}$: $Tb_{0.05}$, $Ce_{0.01}$ phosphors (c) Emission intensities of Tb^{3+} and Ce^{3+} co-doped phosphors as a function of Ce^{3+} concentration, (d) Energy level diagram.

The Ce^{3+} ion is a well-known sensitizer for certain trivalent rare-earth-metal ion luminescence, and the sensitizing effects depend strongly on the host lattices into which these ions are introduced. The $Ca_3Y_2Si_3O_{12}$ phosphor powder is co-doped with cerium and terbium ions, they showed strong green luminescence under UV excitation caused by the line emission of Tb^{3+} ions. The typical excitation and emission spectrum is shown in Fig. 10. The excitation spectrum (Fig. 10(a)) monitored with the 545 nm emission($^5D_4\rightarrow^7F_5$) of Tb^{3+} consists of the excitation bands of the Ce^{3+} and Tb^{3+}ions, indicating that the Tb^{3+} ions are essentially excited through Ce^{3+} ions. Thus energy transfer from Ce^{3+} to Tb^{3+} ions exists in these phosphors. To monitor the green emission (545 nm) of the $Ca_3Y_{1.94}Si_3O_{12}:Tb_{0.05}$, $Ce_{0.01}$ powder phosphor, we used an intense excitation band at 242 nm. From the figure 10(b), we have observed both violet blue emission of Ce^{3+} and intense green emission of Tb^{3+} from the transitions $^5D_4\rightarrow^7F_J$ (J=6,5,4,3). This indicates that an energy transfer from Ce^{3+} to Tb^{3+} occurs in these phosphors. Fig. 10(c) shows the emission intensity of $Ca_3Y_{2-x-y}Si_3O_{12}$: Tb_x,Ce_y phosphors with a fixed Tb^{3+} content (5 mol%) and a varied contents of the co-dopant Ce^{3+} (1, 3 and 5 mol%). An optimum content of cerium was identified to be 1 mol% based on their emission performance. We had observed the enhancement of green emission intensity in the presence of co-dopant (Ce^{3+}) as the sensitizer, upon excitation with an UV-wavelength of 242 nm.

As shown in Fig10(d) the excitation source could first excite Ce^{3+}, subsequently energy transfer takes place from 5d (Ce^{3+}) to the UV absorption levels of Tb^{3+} which decay non-radiatively to the excited level $^5D_4(Tb^{3+})$ from which it decays radiatively down to the various underlying levels of $^5F_{J=0,1,2,3,4,5\&6}$. The energy levels of Tb^{3+} ($4f^n$) are suitable for an energy transfer process to take place from the allowed absorption on the Ce^{3+} upon excitation with an UV source [16, 17]. Absorption of UV by the Ce^{3+} occurs between its ground level ($^2F^{5/2}$ of $4f^1$) and upper level (^2D of 5d). The energy from its return to the ground state is transferred to the Tb^{3+} ion. With increasing Ce^{3+} concentration, the Ce-Tb separation becomes shorter. The higher the Ce^{3+} concentration, the greater the fluorescence quenching as the probability of energy transfer to a second Ce^{3+} increases. The most likely dominant mechanism for the concentration quenching of Ce^{3+} is energy migration to the neighbouring Ce^{3+} ions in their ground state. An energy-transfer probability that changes in proportion with the Ce concentration indicates that the Ce-Tb resonant energy transfer is mainly a dipole interaction. Energy transfer between same rare-earth ions bring about a concentration quenching, while those between different rare-earth ions cause the sensitization of the co-dopant. The electronic configuration of $Tb^{3+}(4f^8)$ is such that it displays several narrow and sharp absorption bands and also sharp and intense emission transitions from the parity-forbidden, weak f-f transitions. The parity allowed transitions (f-d) of Ce^{3+} demonstrate broader absorption bands. This situation induces an efficient green emission from the terbium materials because of the energy transfer processes from Ce^{3+} to Tb^{3+}. The energy level diagram illustrates the energy transfer from Ce^{3+} to Tb^{3+} in a co-doped $Ca_3Y_2Si_3O_{12}$ phosphor upon excitation with an UV source (242 nm) [13-17].

10.2   White light emission of $Ca_3Y_2Si_3O_{12}:Dy^{3+}$ phosphors with Ce^{3+} ion co-doping

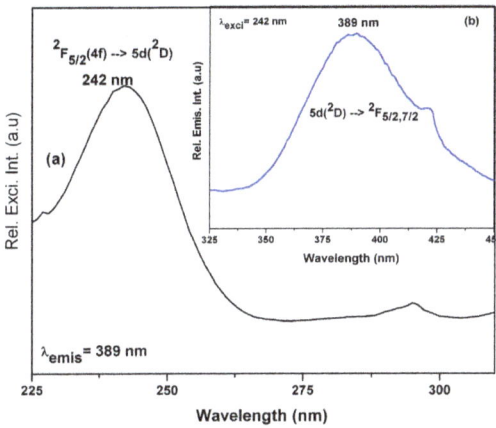

Figure 11. (a) Excitation (b) Emission spectrum of $Ca_3Y_{1.95}Si_3O_{12}$: $Ce_{0.05}$ phosphors.

Fig.11 (a) shows the excitation and emission spectra of $Ca_3Y_{1.95}Si_3O_{12}$: $Ce_{0.05}$ phosphors. The excitation spectrum (Fig.11a) shows a significant band at 242 nm along with a weak band at 295 nm, those bands are associated with the allowed f-d transitions from the ground state $^2F_{5/2}$ to 5d excited level. The emission spectrum (Fig. 11b) consists of two peaks with maxima at 389 nm along with a shoulder peak at 422 nm, corresponding to the transitions from the lowest 5d excited state to the spin-orbit components (^2D) of the doublet ground state, $^2F_{5/2}$ and $^2F_{7/2}$. The energy difference between these two emission peaks is 2010 cm^{-1}, which is basically in accordance with the energy difference obtained by theory between the spin-orbit split $^2F_{5/2}$ and $^2F_{7/2}$ doublets (about 2000 cm^{-1}) [18]. Therefore the Ce^{3+} ion can be used as sensitizer as well as an activator, depending on the splitting of 5d excited levels by the crystal field symmetry.

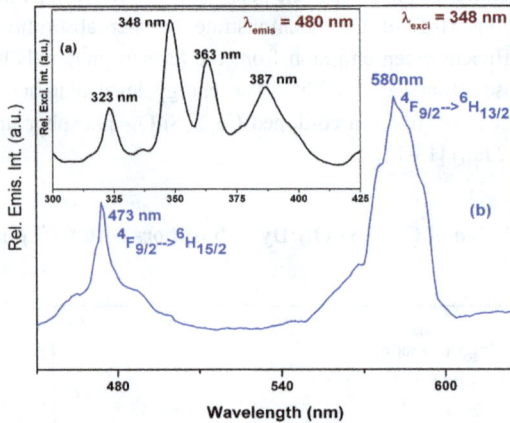

Figure 12. *(a) Excitation (b) Emission spectrum of $Ca_3Y_{1.95}Si_3O_{12}$: $Dy_{0.05}$ phosphors.*

Fig.12a shows an excitation spectrum of Dy^{3+} activated $Ca_3Y_2Si_3O_{12}$ powder phosphor. The excitation spectrum was measured by monitoring the emission at 580 nm, reveals that bands found in the wavelength region of 300-425 nm shows a strong excitation at 348 nm ($^6P_{7/2}$) and also at 323 nm, 363 nm, and 387 nm which correspond to the transitions from the ground state $^6H_{15/2}$ to the excited states $^4K_{15/2}$, $^6P_{5/2}$, $^4I_{13/2}$ respectively. From the measured emission spectrum (Fig.12b), the characteristic transition lines from the lowest excited $^4F_{9/2}$ level of Dy^{3+} to $^6H_{15/2}$ (473 nm) and $^6H_{13/2}$ (580 nm) are observed. The blue ($^4F_{9/2} \rightarrow {}^6H_{15/2}$) emission corresponding to the magnetic dipole transition with selection rule, $\Delta J=1$ and the yellow ($^4F_{9/2} \rightarrow {}^6H_{13/2}$) emission belongs to the hypersensitive transition with the selection rule, $\Delta J=2$. The $^4F_{9/2} \rightarrow {}^6H_{13/2}$ transition is a forced electric dipole transition is only allowed for low symmetries with no inversion centre [19]. When Dy^{3+} is located at low-symmetry local site (without an inversion centre), this emission is often prominent in its emission spectrum. As the radius of Dy^{3+} (1.03 A°) is almost the same as that of Y^{3+} (1.02 A°), so Dy^{3+} can easily enter into the eight-fold coordination Y^{3+} sites of D_{2d} point symmetry (without an inversion centre). Therefore, the ions situated at such low-symmetry local sites for Dy^{3+} ions, the $^4F_{9/2} \rightarrow {}^6H_{13/2}$ transition emission is prominent in the emission spectra. The $(^5D_0 \rightarrow {}^7F_1)/(^5D_0 \rightarrow {}^7F_2)$ emission ratio of Eu^{3+} and $(^4F_{9/2} \rightarrow {}^6H_{15/2})/(^4F_{9/2} \rightarrow {}^6H_{13/2})$ emission ratio of Dy^{3+} exhibit the same trends, with variations in $Ca_3Y_2Si_3O_{12}$. This result proves that the local symmetry of the activator ions belongs to inversion symmetry in the $Ca_3Y_2Si_3O_{12}$ host lattice. However, by observing the Fig.11 and Fig.12 we can state that the emission band of Ce^{3+}-doped sample and excitation band of Dy^{3+}-doped sample overlap in the wide wavelength region from 320 nm and 420 nm. This suggests that there should be energy transfer from Ce^{3+} to Dy^{3+} when co-doped in the same host matrix.

Fig.13 explains the energy transfer mechanism existing between Ce^{3+} and Dy^{3+} ions at room temperature. Figure.13a shows excitation spectrum of co-doped $Ca_3Y_{1.92}Si_3O_{12}$: $Dy_{0.05}$ $Ce_{0.03}$ phosphor. Using λ_{emis} = 580 nm, it is noted that the two intense bands observed in UV region at 242 nm of Ce^{3+} relates to the 4f-5d transition and small peak at 348nm is assigned to the Dy^{3+} ($^6H_{15/2} \rightarrow {}^6P_{7/2}$). The emission spectrum (Fig. 13b) of $Ca_3Y_{1.92}Si_3O_{12}$: $Dy_{0.05}$ $Ce_{0.03}$ phosphor excited at 242 nm consists not only the peaks of Dy^{3+} at 473 nm ($^4F_{9/2} \rightarrow {}^6H_{15/2}$), 580 nm ($^4F_{9/2} \rightarrow {}^6H_{13/2}$), but also the peak of Ce^{3+} at 389 nm (5d-4f).

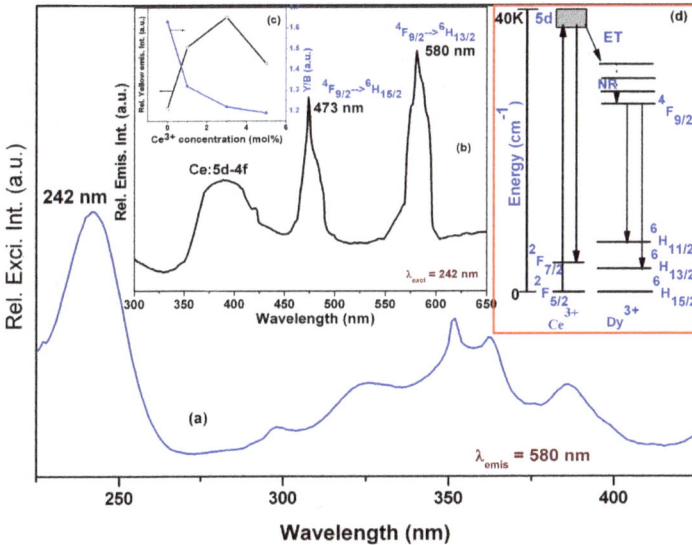

Figure 13. (a) Excitation (b) Emission spectrum of $Ca_3Y_{1.92}Si_3O_{12}$: $Dy_{0.05}$, $Ce_{0.03}$ phosphors (c) Yellow Emission intensities and Y/B intensity ratio of Dy^{3+} and Ce^{3+} co-doped phosphors as a function of Ce^{3+} concentration. (d) Energy level diagram.

If we compare Fig.11 and Fig.12 with Fig.13, a spectral overlap scenario between emission spectrum of Ce^{3+} ion and excitation spectrum of Dy^{3+} ion could be observed.

These significant results indicate that energy transfer may take place from Ce^{3+} to Dy^{3+} [20]. In this present work the concentration of Dy^{3+} ions was fixed at 5 mol% and the concentration of Ce^{3+} ion was varied. Fig. 13c shows the dependence of PL intensity on Ce^{3+} concentration of the yellow band and the yellow-to-blue intensity ratios of $Ca_3Y_{1-x-y}Si_3O_{12}$: $Dy_x Ce_y$ (x = 5 mol%, y = 0, 1, 3 & 5 mol%) phosphors. From the figure 13c we identified that in this system the intensity ratio of Y/B is influenced significantly by the Ce^{3+} concentration. It can be seen that by increasing the Ce^{3+} concentration, the yellow emission at 580 nm of Dy^{3+} increases up to 3 mol% of Ce^{3+}. When the Ce^{3+} concentration is higher than 3 mol%, the emission intensity reduces because of concentration quenching. Also, when the concentration of Ce^{3+} ion is increased, the Y/B intensity ratio's is decreased. The energy level diagram (Fig. 13d) illustrates the energy transfer from Ce^{3+} to Dy^{3+} in a co-doped $Ca_3Y_{1-x-y}Si_3O_{12}$: Dy_xCe_y phosphor upon excitation with an UV source (242 nm)

The excitation peak observed at 242 nm for co-doped phosphor indicates that first, the Ce^{3+} ion is excited from the ground state (4f) to the excited state (5d). In the excited state, this electron either relaxes to the lowest 5d crystal field splitting state then returns to the ground state to produce the blue emission or transfers its excitation energy to the higher excited energy levels of Dy^{3+}, which relax to the $^4F_{9/2}$ level by non-radiation, from where blue ($^4F_{9/2} \rightarrow {}^6H_{15/2}$) and yellow ($^4F_{9/2} \rightarrow {}^6H_{13/2}$) emissions occur. For lower concentrations of Ce^{3+} ions, the energy transfer from Ce^{3+} to Dy^{3+} was efficient and hence the emission performance of Dy^{3+} ions has increased significantly. When the concentration of Ce^{3+} ions was higher, however, the cross relaxation between Ce^{3+} ions led to the slight decrease in the emission intensity of Dy^{3+} ions in the $Ca_3Y_{1.92}Si_3O_{12}$: $Dy_{0.05} Ce_{0.05}$ co-doped phosphor.

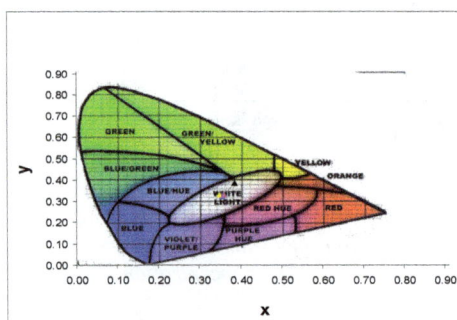

Figure 14. *CIE colour coordinates of $Ca_3Y_{1-x-y}Si_3O_{12}$: Dy_xCe_y (x = 5 mol%, y = 0, 1, 3 & 5 mol%) phosphors.*

As mentioned above, white light can be produced by blending blue and yellow emissions at an appropriate ratio. The Commission International De I- Eclairage (CIE) chromaticity coordinates for $Ca_3Y_{1-x-y}Si_3O_{12}:Dy_xCe_y$ (x = 5 mol%, y = 0, 1, 3 & 5 mol%) phosphors were calculated and are shown in Fig. 14. We have observed that those exhibit excellent CIE coordinates of (0.386, 0.384), (0.355, 0.323), (0.349, 0.33) and (0.36, 0.332) for 0, 1, 3 & 5 mol% of co-dopant (Ce^{3+}) concentration respectively. It is found that the calculated CIE colour coordinates (0.349, 0.33) for $Ca_3Y_{1.92}Si_3O_{12}$: $Dy_{0.05}Ce_{0.03}$ phosphors are very close to the neutral point (0.33, 0.33) [21]. These results affirm that these phosphors are promising materials for producing white light emission.

10.3 Visible-NIR luminescence performance of Nd^{3+} and Er^{3+} ions doped Silver Zinc Borate Host glass [22]

In systems consisting of separate absorbing and emitting components, the absorbing species that can transfer energy to the emitting one is referred to as an antenna [23]. In this situation the excess energy of an excited particle is discarded as a photon in a spontaneous emission process [24]:

$$P^* \rightarrow P + \frac{hc}{\lambda} \tag{1}$$

Where P^* is an excited state. The rate for spontaneous emission in populating a state obeys a first order kinetics law:

$$N_f(t) = N_f(t = 0)\exp(-kt) \tag{2}$$

Where, k is the rate constant for spontaneous emission and defined as $1/\tau$. The value of τ represents the lifetime of the state f [24]. In solids the usual lifetime of a f-f transitions in lanthanides is around 10 µ-1ms [24]. However, the observed lifetime, τ, is dependent on the contribution of both radiative (τ_r) and non-radiative (τ_{nr}) aspects [25]. Therefore, the observed lifetime τ, is only exactly equal to τ_r in the absence of non-radiative decay [24]. This non-radiative deactivation process occurs when the excitation energy is converted into vibrational quanta of the surrounding energies or energy is absorbed by surrounding ions [26]. In fact, the average lifetime for a transition is determined by:

$$\frac{1}{\tau} = \sum_h A_{ab} + \sum_h W_{ab} \tag{3}$$

Where A and W are the probabilities of radiative and non-radiative decay, respectively, for an excited state a [27]. The probability for non-radiative decay is defined by the energy gap law [28]:

$$W = \left(\frac{1}{\tau_0}\right) exp \left(\frac{-a\Delta E}{\hbar\omega_M/2\pi}\right) \qquad (4)$$

Where, the τ_0 and a parameters are fitted empirically. ΔE is the energy gap between the populated state and the next lower lying energy level of a RE ion and ω_M is the highest phonon frequency of the host matrix [28]. From Eq. (4) it is possible to verify that W increases for small energy gaps. Thus, when the energy gaps between the different levels are too small W is high and luminescence processes do not occur. In this situation the energy is transferred by a multiphonon process. Usually, in glass samples $W>A$, resulting in a low quantum efficiency [29]. In fact, a separation of at least five phonons is required to observe radiative processes [25]. This concept is described in Fig. 15. The energetic levels were assigned accordingly to the values presented in reference [30].

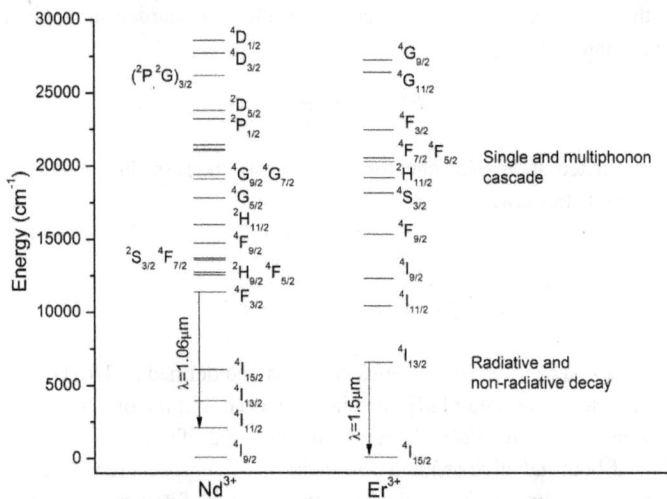

Figure 15. Energy level diagram of Nd^{3+}and Er^{3+}: silver zinc borate glasses [22].

In this figure are described the *f-f* transitions allowed for Nd^{3+} and Er^{3+}. It is possible to distinguish two main ranges of energy where different physical processes can occur. For higher energy levels, the energy gap between two consecutive states is very small. Thus, the electronic transitions in this region are most of the time governed by non-radiative decay processes. When the gap between energy levels increases it is possible to observe strong luminescent effects, because W decreases. Nevertheless, non-radiative deactivation can occur. In fact, this process is not well understood. There are many theories explaining the mechanism of non-radiative decay, which are of major importance in laser applications, phosphorus luminescence as discussed in the above sections etc [31].

Fig. 16 shows the visible emission spectra for both, Nd^{3+} and Er^{3+} doped silver lead borate glasses. Some of the principal energy bands are well discriminated in the image. However these bands are broad, instead of being sharp. Usually the positions of the peaks only vary slightly, if at all, with the environment. However their intensities are strongly dependent on the type of glass matrix [32]. Moreover, the sites and environment experienced by the ions are not homogeneous in the glass matrix, which results in a inhomogeneous broadening of spectral features [25]. In the same fashion, this lack of symmetry and homogeneous distribution of the luminescent centres will also have an impact in the time-resolved decay kinetics.

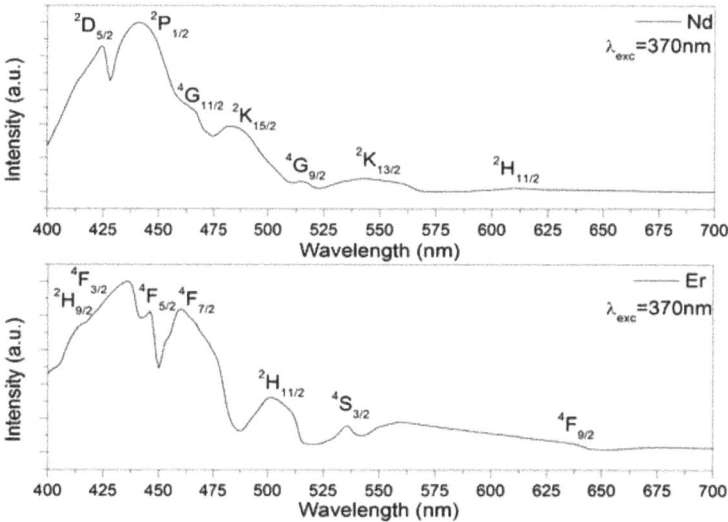

Figure 16. Visible luminescence spectra of Nd^{3+} and Er^{3+}: Silver zinc borate glasses [22].

In the NIR, neodymium doped all-solid-state laser sources have been recognized as the most efficient laser sources for numerous applications in the fields of electronics, optics and communications. The NIR transition at 1050 nm ($^{4}F_{3/2} \rightarrow ^{4}I_{11/2}$) made the Nd-glass laser the most common laser on the market [33-35]. In the same way, the Erbium emission peak at around 1550 nm ($^{4}I_{13/2} \rightarrow ^{4}I_{15/2}$) has attracted a lot of attention in the past few years because of possible applications in optical communication systems and optical switching [33]. Fig. 17 shows the NIR luminescence spectra of of Nd^{3+} and Er^{3+} doped silver zinc borate glasses. The emission peaks are clearly identifiable, but in order to be used in optics or other applications, the glasses need more detailed characterization. Table 3 provides some features of the two NIR transitions for Nd^{3+} and Er^{3+} along with relevant literature data.

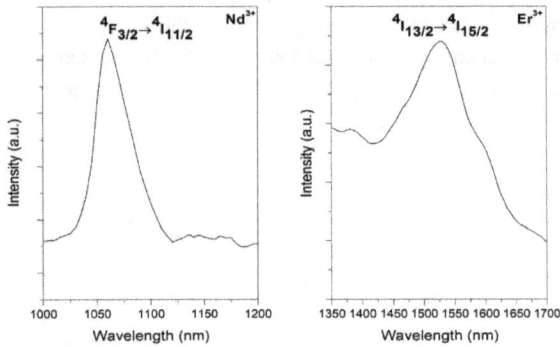

Figure 17. NIR luminescence spectra of Nd^{3+} and Er^{3+}: Silver zinc borate glasses[22].

Table 3: *NIR transitions emission parameters of Nd^{3+} and Er^{3+}: Silver zinc borate glasses along with literature data [22].*

Glasses	λ_{exc} (nm)	NIR Transition	λ_{emi} (nm)	FWHM (nm)	τ_{ave} (µs)	Stimulated emission cross-sections $(\sigma^P_e X10^{-19} cm^2)$	FOM FWHM X σ^P_e (nmX10^{-19}cm^2)
(1)	(2)	(3)	(4)	(5)	(6)	(7)	(8)
Nd$_2$O$_3$ (Present)	370	$^4F_{3/2} \rightarrow {}^4I_{11/2}$	1060	41	46	3.36	137.76
Literature data							
20TeO$_2$-1Nd$_2$O$_3$[38]	495	$^4F_{3/2} \rightarrow {}^4I_{11/2}$	1065	29.97	35	0.27	--
B$_2$O$_3$[39]	514	$^4F_{3/2} \rightarrow {}^4I_{11/2}$	1060	57.2	50.5	0.50	--
Fluorophosphate[35]	800	$^4F_{3/2} \rightarrow {}^4I_{11/2}$	1058	32.00	358	0.27	--
Er$_2$O$_3$ (Present)	970	$^4I_{13/2} \rightarrow {}^4I_{15/2}$	1530	101	255	1.04	105.04
Literature data							
20TeO$_2$-1Er$_2$O$_3$[36]	970	$^4I_{13/2} \rightarrow {}^4I_{15/2}$	1532	67	446	0.90	60.30
	980	$^4I_{13/2} \rightarrow {}^4I_{15/2}$	1550	75	--	0.70	52.72
YB25[23]	970	$^4I_{13/2} \rightarrow {}^4I_{15/2}$	1550	79	--	0.70	55.40

The full width at half maximum (FWHM) and stimulated emission cross-section (σ^E_p) are of major importance in optics. The FWHM can be directly measured from the spectra, however σ^E_p must be calculated by [36]:

$$\sigma^E_p = \left(\frac{\lambda_p^4}{8\pi c n_d^2 \Delta\lambda}\right) A \qquad (5)$$

where λ_p is the wavelength of the emission peak and $\Delta\lambda$ the effective half-width of the emission band, c is the velocity of the light, n_d is the refractive index of the glass and A is the transition probability. The values obtained (Table 3, column 7) are in good agreement with literature data, thus revealing the quality of the prepared glasses. The quality of a glass can also be evaluated by determining Figure of Merit (FOM) (FWHM x σ^E_p). In fact, the gain bandwidth of an amplifier is determined by FOM and acts as an indicator of better quality [37]. Although, not many references for FOM are found in the literature, it is possible that our results (Table 3 column 8) are indicative of a good quality glass.

Xiang Shen [40] *et al.* have discussed the energy transfer (ET) mechanism between Er^{3+} and Nd^{3+} on their energy level characteristics. The interaction parameters, C_{D-A}, for the energy transfer (ET) processes from Er^{3+} to Nd^{3+} in tellurites glass was calculated. In the present borate host glasses, the luminescence was characterized using steady state techniques and the strongest NIR emission peaks observed were $^4F_{3/2} \rightarrow {}^4I_{11/2}$ for Nd^{3+} and $^4I_{13/2} \rightarrow {}^4I_{15/2}$ for Er^{3+} ions respectively.

11. Conclusions

In summary, the Ce^{3+} and/or Tb^{3+}, Dy^{3+} -doped $Ca_3Y_2Si_3O_{12}$ novel phosphors have been synthesized successfully by a sol-gel method with the citric acid as the chelating agent in precursor solutions. The Ce^{3+} excitation bands are due to the transitions from 4f level to the crystal field splitting levels of 5d. The Tb^{3+} excitation bands are ascribed to absorption of the spin-forbidden and spin-allowed transitions from 4f to 5d state of the Tb^{3+} ion. Upon UV excitation, the doped rare-earth ions showed their characteristic emission, i.e., Ce^{3+} 5d-4f (Blue) and Tb^{3+} $^5D_{3,4} \rightarrow {}^7F_J$ (Green) transitions respectively. PL excitation and emission spectra demonstrated that there exits an efficient energy transfer from Ce^{3+} to Tb^{3+} in the co-doped phosphors which resulted in enhanced green emission from it.

The PL properties measured at room temperature under UV excitation clearly confirm the characteristic emission features of Ce^{3+} (5d-4f) and Dy^{3+} ($^4F_{9/2} \rightarrow {}^6H_{15/2}$ and $^4F_{9/2} \rightarrow {}^6H_{13/2}$) transitions respectively. Furthermore, we have observed that with an increase in Ce^{3+} concentration, the emission associated with Dy^{3+} increased greatly up to 3 mol% of Ce concentration, which indicated there exists an energy transfer from Ce^{3+} to Dy^{3+} in $Ca_3Y_2Si_3O_{12}$. This provides an opportunity to identify a novel phosphor for obtaining white light emissions from a single luminescent system. Furthermore, the energy transfer phenomena of co-doped Er^{3+} to Nd^{3+} silver zinc borate host system is under development stage.

12. Acknowledgement

One of the Author (GBK) would like to thank the Department of Science and Technology (DST) and Science and Engineering Research Board (SERB) for granting a project under Young Scientist in Engineering Sciences scheme.

This work is dedicated to the beloved teacher late Professor Srinivasa Buddhudu for his continuous encouragement with helpful discussions and good advice to his PhD students namely; BCB, BVR, GBK and NSH.

References

[1]   C. C. Lin and R. S. Liu, Advances in Phosphors for Light-emitting Diodes, J. Phys. Chem. Lett. 2011; 2: 1268-1277. http://dx.doi.org/10.1021/jz2002452

[2]   S. Ye, F. Xiao and Y.X. Pan, Phosphors in phosphor-converted white light-emitting diodes: Recent advances in materials, techniques and properties, Mater. Sci. Engg. R (2010); 71:1-43. http://dx.doi.org/10.1016/j.mser.2010.07.001

[3]   G Li, D Geng, M Shang, Color Tuning Luminescence of $Ce^{3+}/Mn^{2+}/Tb^{3+}$-Triactivated $Mg_2Y_8(SiO_4)_6O_2$ via Energy Transfer: Potential Single-Phase White-Light-Emitting Phosphors, J. Phys. Chem.2011; C 11: 21882–21892.

[4]   H A. Hoppe, Recent Developments in the Field of Inorganic Phosphors, Angewandte Chemie International Edition 2009;48: 3572. http://dx.doi.org/10.1002/anie.200804005

[5]   H.W. Leverenz, An introduction to luminescence of solids, John Wiley & Sons, 1950, 399.

[6]   G. Blasse and B.C. Grabmaier , Luminescent materials, Springer Verlag, Berlin, 1994. http://dx.doi.org/10.1007/978-3-642-79017-1

[7]   W.M. Yen, S.Shionoya, H.Yamamoto, Phosphor handbook, CRC Press, Taylor & Francis Group, 2007.

[8]   V.B. Pawade, H.C. Swart, S.J. Dhoble, Review of rare earth activated blue emission phosphors prepared by combustion synthesis, Renew. Sustainable Energy Rev. 2015; 52: 596–612. http://dx.doi.org/10.1016/j.rser.2015.07.170

[9]   See, http://web.mit.edu/is08/pdf/Lighting%20to%20distribute%2010Apr2008.pdf

[10] See, http://img.ledsmagazine.com/pdf/LightingtheWay.pdf

[11] See, http://hyperphysics.phy-astr.gsu.edu/hbase/vision/cie.html

[12] E. Pavitra, G. Seeta Rama Raju, Yeong Hwan Ko, and Jae Su Yu, Novel strategy for the controllable emissions from Eu^{3+} or Sm^{3+} ions co-doped SrY_2O_4: Tb^{3+} phosphor, Phys Chem Chem Phys. 2012;14(32):11296-307. http://dx.doi.org/10.1039/c2cp41722g

[13] H. Lai, A. Bao, Y. Yang, Y.Tao, H.Yang, Y. Zhang, and L. Han, UV Luminescence Property of YPO_4:RE (RE = Ce^{3+}, Tb^{3+}), J. Phys. Chem. C, 2008;112: 282-286. http://dx.doi.org/10.1021/jp074103g

[14] D. Jia, R. S. Meltzer, W. M. Yen, Green phosphorescence of $CaAl_2O_4$: Tb^{3+},Ce^{3+} through persistence energy transfer, Appl. Phys. Lett., 2002; 80: 1535-37. http://dx.doi.org/10.1063/1.1456955

[15] B.V. Rao, I.Chan, Effect of calcination temperature and concentration on luminescence properties of novel Ca3Y2Si3O12: Eu phosphors, J. Amer. Ceram. Soc. 2009;92(12):2953–2956. http://dx.doi.org/10.1111/j.1551-2916.2009.03308.x

[16]   P. Dorenbos, The 4f n↔4f n-15d transitions of the trivalent lanthanides in halogenides and chalcogenides, J. Lumin.2000; 91: 91-106. http://dx.doi.org/10.1016/S0022-2313(00)00197-6

[17]   J.M.F. van Dijk and M.F.H.Schuurmans, On the nonradiative and radiative decay rates and a modified exponential energy gap law for 4f-4f transitions in rare-earth ions, J.Chem. Phys. 1983;78(9):5317-5323. http://dx.doi.org/10.1063/1.445485

[18]   S. Erdei , F.W. Ainger, D. Ravichandran, W.B. White, L.E. Cross, Preparation of Eu^{3+} :YVO_4, red and Ce^{3+}, Tb^{3+} : $LaPO_4$ green phosphors by hydrolyzed colloid reaction (HCR) technique, Mater. Lett. 1997;30: 389-393. http://dx.doi.org/10.1016/S0167-577X(96)00230-3

[19]   G. Bhaskar Kumar, S. Buddhudu, Synthesis and emission analysis of RE^{3+}(Eu^{3+} or Dy^{3+}) : Li_2TiO_3 ceramics, Ceram. Int. 2009;35: 521-525. http://dx.doi.org/10.1016/j.ceramint.2007.09.107

[20]   J. Lin, Q. Su, Luminescence and Energy Transfer of Rare-earth-metal ions in $Mg_2Y_8(SiO_4)_6O_2$, J. Mater. Chem. 1995; 5(8): 1151-1154. http://dx.doi.org/10.1039/jm9950501151

[21]   Z. Lei, G.Xia, L. Ting, G.Xiaoling, L.Ming, S.Guangdi, Color rendering and luinous efficacy of trichromatic and tetrachromatic LED-based white LEDs, Microelec. J, 2007; 38: 1-6. http://dx.doi.org/10.1016/j.mejo.2006.09.004

[22]   J. Coelho, G. Hungerford and N. Sooraj Hussain, Luminescence and time-resolved emission spectra of Nd^{3+} and Er^{3+}: Silver zinc borate glasses, Solid State Phenom 2014; 207:37-54. http://dx.doi.org/10.4028/www.scientific.net/SSP.207.37

[23]   F. Artizzu, ML. Mercuri, et.al., NIR-emissive Erbium Quinolinolate Complexes. Coord Chem Rev. 2011. http://dx.doi.org/10.1016/j.ccr.2011.01.013

[24]   M. Gaft M, R. Reisfeld, G. Panczer, Luminescence spectroscopy of minerals and materials: Springer Verlag; 2005.

[25]   Hänninen P, Härmä H. Lanthanide Luminescence: Photophysical, Analytical and Biological Aspects: Springer; 2011. http://dx.doi.org/10.1007/978-3-642-21023-5

[26]   R. Reisfeld . Radiative and non-radiative transitions of rare-earth ions in glasses. Rare Earths. 1975:123-75. http://dx.doi.org/10.1007/bfb0116557

[27]   J. Coelho J, G.Hungerford , N.S. Hussain. Structural and time resolved emission spectra of Er^{3+}: Silver lead borate glass. Chem Phys Lett. 2011;512:70-5. http://dx.doi.org/10.1016/j.cplett.2011.07.019

[28]   BZ. Malkin, eta.al., Theoretical studies of nonradiative 4f–4f multiphonon transitions in dielectric crystals containing rare earth ions. J Mol Struct. 2007;838:170-5. http://dx.doi.org/10.1016/j.molstruc.2007.01.009

[29] C. Layne, W. Lowdermilk, MJ. Weber. Multiphonon relaxation of rare-earth ions in oxide glasses. Phys Rev B: Condens Matter. 1977;16:10. http://dx.doi.org/10.1103/PhysRevB.16.10

[30] W. Carnall, P. Fields, K. Rajnak, Electronic energy levels in the trivalent lanthanide aquo ions. I. Pr, Nd, Pm, Sm, Dy, Ho, Er, and Tm. J Chem Phys. 1968;49:4424. http://dx.doi.org/10.1063/1.1669893

[31] E. Sveshnikova, V. Ermolaev, Inductive-resonant theory of nonradiative transitions in lanthanide and transition metal ions (review). Optics and Spectroscopy. 2011;111:34-50. http://dx.doi.org/10.1134/S0030400X11070186

[32] R. Reisfeld, T. Saraidarov, E. Ziganski, et.al, Intensification of rare earths luminescence in glasses. J Lumin. 2003;102–103:243-7. http://dx.doi.org/10.1016/S0022-2313(02)00506-9

[33] R. Reisfeld., Optical Properties of Rare Earth and Transition Element Doped Glasses. In: Editors-in-Chief: KHJB, Robert WC, Merton CF, Bernard I, Edward JK, Subhash M, et al., editors. Encyclopedia of Materials: Science and Technology (Second Edition). Oxford: Elsevier; 2001. p. 6472-7. http://dx.doi.org/10.1016/b0-08-043152-6/01145-1

[34] R. Rajeswari, S. Babu, CK. Jayasankar, Spectroscopic characterization of alkali modified zinc-tellurite glasses doped with neodymium. Spectrochim Acta, Part A. 2010;77:135-40. http://dx.doi.org/10.1016/j.saa.2010.04.040

[35] JH.Choi, A.Margaryan, et.al, Judd–Ofelt analysis of spectroscopic properties of Nd^{3+}-doped novel fluorophosphate glass. J Lumin. 2005;114:167-77. http://dx.doi.org/10.1016/j.jlumin.2004.12.015

[36] J. Coelho, J. Azevedo, G.Hungerford, NS. Hussain, Luminescence and decay trends for NIR transition ($^4I_{13/2} \rightarrow {}^4I_{15/2}$) at 1.5µm in Er^{3+}-doped LBT glasses. Opt Mater. 2011;33:1167-73. http://dx.doi.org/10.1016/j.optmat.2011.02.003

[37] H. Fan, G. Wang, K. Li, Broadband 1.5- emission of high erbium-doped Bi_2O_3–B_2O_3–Ga_2O_3 glasses. Solid State Commun. 2010;150:1101-3. http://dx.doi.org/10.1016/j.ssc.2010.03.031

[38] J. Yang, S. Dai, et.al., Spectroscopic properties and thermal stability of erbium-doped bismuth-based glass for optical amplifier. J. Appl. Phys. 2003;93:977-83. http://dx.doi.org/10.1063/1.1531840

[39] B. Karthikeyan, S. Mohan. Spectroscopic and glass transition investigations on Nd^{3+}-doped NaF–Na_2O–B_2O_3 glasses. Mater Res Bull. 2004;39:1507-15. http://dx.doi.org/10.1016/j.materresbull.2004.04.025

[40] Shen Xiang, Qiuhua Nie, et al., Investigation on energy transfer from Er^{3+} to Nd^{3+} in tellurite glass, J. Rare Earths, 2008; 26: 899-903. http://dx.doi.org/10.1016/S1002-0721(09)60029-6

Keywords

About the Editor

Dr. Sooraj Hussain Nandyala

Dr Sooraj employed at the University of Birmingham as a project manager in the Marie Skłodowska-Curie, (Research and Innovation Staff Exchange) project. His important contribution is to develop innovative advanced composite materials for antimicrobial, biomechanical, bone regeneration and coating applications. He has worked on luminescence and time-resolved emission spectroscopy studies of different lanthanide host glasses along with 3D printing technology for next generation of biomaterials. Dr Sooraj has demonstrated his ability to work in synergistic multidisciplinary aspects. He is very active and an independent member of the biomaterials research community proven by his track record, editorial activities and leadership in the EU cost actions. The Portuguese Ministry of Science & Technology (FCT) awarded him with one major nation-wide project, and two bilateral joint collaborative projects with the Ministry of Science & Technology, India. He also completed several industrial consultancy projects. He has supervised several masters, project fellows and postdoctoral. He has 8 books to his credit and has published more than 70 research papers and 8 book chapters with a total number of 908 citations, h-index 18 (Google scholar database). He is one of the core group members of technologies for Optofluidic devices and working group leader for materials (soft, bio & nano) in the EU Cost Action MP1205 and participated in 8 EU Cost Action Meetings (Belgium, Czech Republic, Ireland, Italy, Portugal, and Spain). Dr Sooraj has also been an external examiner for several Ph.D. theses in Portugal and abroad and has reviewed several research proposals and CRC press books. He is acting as an Editor in Chief of the Journal of Biomimetics, Biomaterials, and Biomedical Engineering - JBBBE, Switzerland. He has significant experience in working with different reputable international laboratories & participating in the European Project POLARIS funded by FP7 at 3B's (Biomaterials, Biodegradable, and Biomimetics) research group in the headquarters of the European Institute of Excellence on Tissue Engineering and Regenerative Medicine, University of Minho, Portugal. He graduated with on M.Sc in Physics, 1995, M.Tech in Energy Management, 1997, and Doctorate in Physics (PhD, 2002) from Sri Venkateswara University, Tirupati, India. In the period (2002-2008), he worked as a postdoctoral fellow in the Instituto de Engenharia Biomédica (INEB) and then he was employed (2008-2014) by INESC Porto, University of Porto, Portugal. He also gained industrial experience with Biosckin S.A. Portugal. And fellow member of International Congress of Chemistry and Environment-FICCE, India.

www.ingramcontent.com/pod-product-compliance
Lightning Source LLC
Chambersburg PA
CBHW071213210326
41597CB00016B/1796